Object Relationship Notation (ORN) for Database Applications

Enhancing the Modeling and Implementation of Associations

Object Relationship Notation (ORN) for Database Applications

Enhancing the Modeling and Implementation of Associations

by

Bryon K. Ehlmann
Southern Illinois University Edwardsville
Edwardsville, IL, USA

Bryon K. Ehlmann
Department of Computer Science
Southern Illinois University Edwardsville
Edwardsville, IL 62026
behlman@siue.edu

ISBN 978-1-4419-3493-2 e-ISBN 978-0-387-09554-7
DOI 10.1007/978-0-387-09554-7
Springer Dordrecht Heidelberg London New York

Printed on acid-free paper

Springer is part of Springer Science+Business Media (www.springer.com)

To Barbara,
Bethany and Jonathan

Preface

Conceptually, information that is relevant to a database application consists of objects and relationships. Since the advent of object-orientation in the 1980s, much emphasis has been placed on objects and, in my opinion, too little on the relationships that exist among them. Many books have been written about objects. This book is about relationships.

The emphasis on objects in software systems analysis, design, and implementation has improved the productivity of developing and maintaining software systems. I believe that placing more emphasis on relationships can have a similar effect, especially for database systems. This book is founded on that belief.

Motivation

The motivation for this book is the same as the motivation driving the research and its results that are the subject of this book. This motivation originated in 1990 when I attended a class on Ontos, an early object database management system (ODBMS). Despite much support in the system for the storing of objects, i.e., object persistence, I saw little support for relationships, specifically the types of relationships that bind different objects together. A classic example of this type of relationship is the one-to-many relationship that binds each object representing a department to the objects representing the employees who work for the department. The support for such relationships improved little in later ODBMSs and in a subsequent ODBMS standard. I viewed the support for relationships in ODBMSs to be slightly less than that provided by relational database management systems (RDBMSs).

And this latter support has, I believe, been dismal, especially when one considers the age and pervasiveness of RDBMSs and the fact that for decades relational databases have been developed based on the Entity-Relationship (ER) model. This belief has provided even more motivation for the contents of this book. When the relationships defined in ER models are implemented in relational databases, much of their meaning—expressed by their one-to-one, one-to-many, many-to-many, and cardinality notations—is lost or very difficult to resurrect. Support for these concepts is lacking in RDBMSs, which makes implementation unnecessarily more difficult and error-prone.

I will "come clean" on another motivation for my research and this book. I confess I have a passion for relationships. I find them fascinating! There are so many different kinds, and it is often a challenge to try to discover their essence. What is it that makes one type of relationship different from or similar to another? Some aspects of a relationship's nature can be defined mathematically, but others often seem only "vague notions." For example, the notions of "containment" contribute to the nature of many relationships, e.g., a car contains, or is composed of, many parts. But what exactly is containment? It seems to come in different "flavors"—e.g., a car also contains materials like plastic, which aren't really "removable," and a car may contain passengers, which certainly must be removable. Can these flavors of containment somehow be generically defined and distinguished so that they can be ultimately and properly "appreciated," i.e., managed, by a DBMS?

Purpose

The purpose of this book is to highlight in more detail the lack of support for relationships in DBMSs but, more importantly, to point a way toward improvement. To this end, I describe in this book a simple, yet powerful notation that modestly extends the basic ER model and its more modern, standard incarnation, the UML class diagram. This notation, the Object Relationship Notation (ORN), allows the true nature of relationships—more specifically, "associations" using UML terminology—to be more precisely defined. It also allows these relationship definitions to be included in a data definition language (DDL), like SQL. This permits a more direct mapping from model to implementation and facilitates better support for relationships in DBMSs, both object and relational.

In this book, I also describe patterns and software tools that demonstrate how ORN can be used to more productively model and implement databases. The patterns are given in Chapter 4, and the software tools are available via the Web. (See the Downloads section later in this Preface.) The patterns, called *association patterns*, assist in developing a better understanding of relationships and in modeling data, regardless of whether ORN is used. The software tools—ORN Simulator, Object Relater *Plus*, and ORN Additive—can be used as research tools to verify the examples given in this book, to test others, and to serve as prototypes for development efforts that would integrate ORN into a commercial modeling tool or DBMS. In addition, the software tools can be used to assist in the development of real database applications, and the ORN Simulator can serve as a pedagogical tool for learning the concepts of data modeling and transaction processing.

The research results I describe in this book do not relate to a "sizzling" area of research in computer science. Rather, they relate to a traditional area—namely, data modeling and database definition—where little research is occurring, where the major advancements were made in the 1970's, where much effort since then has been on standardization, and thus where any breakthrough is slow to be adopted. Nevertheless, this book is an effort on my part to present and promote some practical research results that I feel can advance the state-of-the-art in this traditional area. I believe that these results can be adopted with little cost and can significantly improve the productivity of developing database systems and improve their integrity.

Readers

The intended readers for this book are researchers in database systems, developers of DBMSs or data modeling tools, practitioners of database systems development, and students of database management. Others who may be interested are software engineering researchers or anyone having an interest in data modeling, database development, or simply learning more about relationships. The book can be used as a supplemental text for courses in database management or database modeling where students use association patterns or one or more of the tools that are discussed.

The prerequisite knowledge for the reader is a basic understanding of data structures, files, and databases. The database knowledge needed is that normally obtained in an introductory course in database management—most importantly, a familiarity with the ER model, relational databases and SQL, and to a lesser extent object databases. In Chapter 1, I provide a brief overview of all of these topics. I use UML class diagrams extensively throughout the book. In Chapter 1, I explain these diagrams so that a reader familiar with ER diagrams should develop a sufficient understanding.

Structure

This book is divided into three parts. The first part contains introductory material about relationships in general and ORN in particular. This material is intended for all readers. The second part contains material for readers interested in using ORN to develop database applications or for readers merely interested in developing a better understanding of the benefits and capabilities of ORN. The third part is for readers interested in including support for ORN in database modeling tools or DBMSs or for readers merely interested in investigating the algorithms required for implementing ORN. Here is a brief summary of each chapter:

Part I About Relationships and ORN

- Chapter 1, **Introduction: Including a Brief History of Relationships**, provides a general introduction to relationships, gives a brief history of relationships in terms of how we have understood and recorded them over the years, and discusses some of the problems encountered today in modeling and recording, i.e., implementing, them using DBMSs. The historical account provides a short review of relational databases and UML class diagrams, which should prove helpful in understanding the remaining contents of the book.
- Chapter 2, **Object Relationship Notation (ORN)**, provides the syntax and semantics for ORN—i.e., its form, both textual and graphical, and its meaning—and gives examples of associations to illustrate the meaning of the notation's symbols.
- Chapter 3, **ORN Simulator: A Modeling Tool Where Associations Come Alive**, presents a modeling tool that the reader can access via the Web. This tool allows its user to easily experiment with ORN and thus develop a better under-

standing of its semantics. The user can easily create an ORN-extended database model and then manipulate a prototype database in the context of the model. By creating and deleting objects and creating, destroying, and changing associations in the database, the user readily sees the effects of different ORN specifications.
- Chapter 4, **Association Patterns: Emerging from a Variety of Association Types**, provides examples of the variety of associations that can be defined using ORN and identifies among them some association patterns that can be used to guide database modeling.
- Chapter 5, **Comparing ORN to Similar Declarative Schemes**, compares ORN to similar schemes for declaring relationship semantics in terms of simplicity and expressive power. It also discusses how ORN relates to efforts to better define the semantics of whole-part relationships.

Part II Using ORN to Develop a Database System

- Chapter 6, **ORN Additive: A Tool for Extending SQL Server with ORN**, discusses a tool that allows its user to add ORN support to Microsoft's SQL Server. The tool automatically generates the T-SQL triggers and stored procedures that are required to implement ORN-defined associations in a database application.
- Chapter 7, **Object Relater *Plus* (OR+): An ORN-Extended Object DBMS**, discusses a tool that allows its user to add ORN support to Progress's Object Store. The tool automatically generates the C++ methods that are required to implement ORN-defined associations in an object database application.
- Chapter 8, **Mapping Database Models to DDLs: From ORN-Extended Class Diagrams to ORN-Extended DBMSs**, shows how one can easily map the ORN-defined associations in a class diagram to an ORN Additive/T-SQL definition of a database or an OR+ Object Database Definition Language (ODDL) definition of a database.
- Chapter 9, **Association Semantics: Dealing with the Subtleties, Inconsistencies, and Ambiguities**, discusses some of the finer points about association semantics and how certain association definitions can lead to associations and combinations of associations that are mathematically inconsistent, likely inconsistent, or ambiguous. Database developers can better identify and deal with such associations when their semantics are defined by ORN.

Part III Adding ORN to the DBMS

- Chapter 10, **A Conceptual Implementation of ORN: Exploring Semantic Circularity and Ambiguity**, provides algorithms, which are independent of database type, for the implementation of ORN. Based on these algorithms, it also discusses the circularity and clarity of ORN semantics in the presence of link cycles within a database. A theorem is stated and proved about the clarity of ORN semantics.
- Chapter 11, **Adding ORN to the SQL Standard for RDBMSs**, provides the syntax and semantics for adding ORN to the SQL relational DBMS standard.
- Chapter 12, **Adding ORN to the ODMG Standard for ODMSs**, provides the syntax, semantics, and algorithms for adding ORN to the ODMG 3.0 standard for Object Data Management Systems (ODMSs).

The diagram in Fig. P.1 shows the dependencies between chapters and thus the order in which chapters can be read. A reader very knowledgeable of data modeling, class diagrams, and database management systems can skip Sections 1.1 and 1.2 of Chapter 1, but should read Sections 1.3 and 1.4 for a proper introduction to the subject matter of this book. If desired, Chapters 2 and 3 can be studied together to allow experimentation and perhaps a better understanding of the contents of Chapter 2.

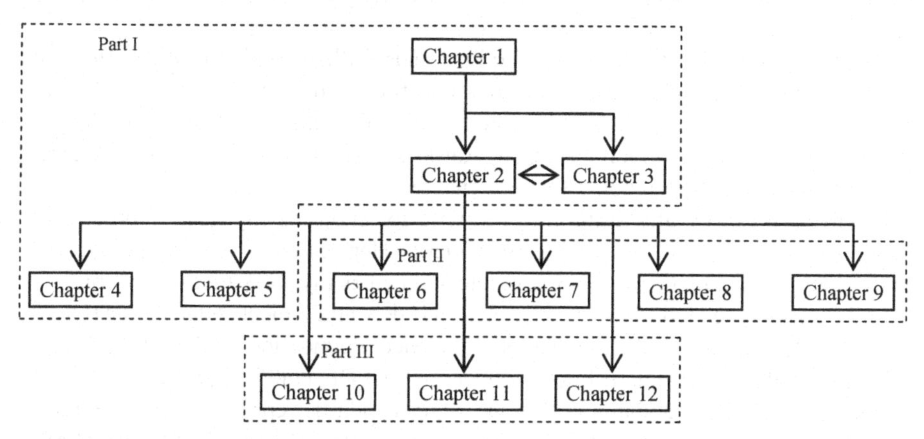

Fig. P.1. Dependencies between chapters

Downloads

The ORN Additive and Object Relater *Plus* (OR+) software tools, which are discussed in this book, can be downloaded from www.siue.edu/~behlman. Access the site and click on the "Download Software" link.

Acknowledgments

A number of people and organizations have contributed to the writing of this book and the work that produced its contents.

I thank the Computer Science Department and School of Engineering at Southern Illinois University Edwardsville and the Computer and Information Systems Department at Florida A&M University for their support. I particularly thank the students in my DBMS, software engineering, and independent study classes for the projects they completed over the course of many semesters that resulted in the extensive testing of the ORN Simulator, OR+, and ORN Additive software prototypes and the association patterns presented in Chapter 4. Special recognition in this regard should go to Anwesha Chattopadhyay, Brock Green, Jian Li, and Rajesh Vaddi.

I thank the co-authors on my previous papers related to ORN for their reviews and assistance. They are Larry Dennis, Siebert Hardeman, Erika Neal, Gregory Riccardi, Naphtali Rishe, Michael Stewart, Jinyu Shi, and Xudong Yu. Gregory Riccardi deserves special recognition for giving me the latitude and support that made my early efforts in developing ORN possible.

I thank my editor at Springer, Susan Lagerstrom-Fife, and especially her assistant, Sharon Palleschi, for working with me, reviewing my drafts, and answering my many questions.

I thank my wife Barbara Ehlmann and Professor William White for their efforts in reviewing and proofreading the entire book and finding my many errors. Still more errors were found by the much appreciated efforts of Bethany Ehlmann, my daughter, and three students—Gerald Jackson, Aaron Jansen, and John Jenkins—in carefully reviewing and proofreading selected chapters.

Finally, I am especially grateful to my wife for enduring my many moments of mental preoccupation while writing this book. Her patience, understanding, and tremendous support made this book possible and makes my work more enjoyable.

Early research related to the development of ORN and Object Relater *Plus* (OR+) was partially supported by the National Science Foundation under grant CDA-9313299 and the U.S. Dept. of Energy under contract DE-FG05-92ER40735.

Jim Melton, SQL standards editor and SQL standards representative for Oracle Corporation, provided helpful clarification on referential integrity for SQL:2003 reference types, which was relevant to Chapter 10.

The opening quotation in Chapter 1 is extracted with permission from (Zdonik and Maier 1990), © 1990, Morgan Kaufmann.

Edwardsville, Illinois Bryon K. Ehlmann

About the Author

Bryon K. Ehlmann is currently Professor of Computer Science at Southern Illinois University Edwardsville. He earned his B.S. and M.S. degrees in Computer Science from the University of Missouri at Rolla (now the Missouri University of Science and Technology) and his Ph.D. degree in Computer and Information Sciences from Florida State University. Professor Ehlmann's research interests include database management, object databases, data modeling, software engineering, and user interfaces. He has twelve years of industry experience with Burrough's Corp. and Unisys Corp. in developing software systems in the areas of database management, report generation, and user interface development. In addition, he has twenty-seven years of academic experience in teaching programming, data management, and software engineering and has published over twenty-five papers in the areas of database management and data modeling. Professor Ehlmann can be contacted at behlman@siue.edu and would appreciate any comments, questions, or corrections related to this book.

Contents

List of Abbreviations

API	application program interface
DBMS	database management system
DBTG	Database Task Group
DDL	data definition language
ER	Entity-Relationship
MDD	model-driven development
ODBMS	object database management system
ODDL	Object Database Definition Language
ODL	Object Definition Language
ODM	object-to-database mapping
ODMG	Object Data Management Group
ODML	Object Database Manipulation Language
ODMS	object data management system
OML	Object Manipulation Language
OO	object-oriented
OQL	Object Query Language
OR+	Object Relater *Plus*
ORN	Object Relationship Notation
RDBMS	relational database management system
RXC	Relationship eXChange
SQL	Structured Query Language
UML	Unified Modeling Language

Part I

About Relationships and ORN

Chapter 1

Introduction

Including a Brief History of Relationships

> A relationship is a named correspondence between objects. Much work in data modeling has focused on understanding the many types of relationships that naturally surface in any application. Relationships are one of the most fundamental parts of any data model. From one point of view, they are what distinguishes databases from file systems.
>
> (Zdonik and Maier 1990, p. 21)

Relationships are even more fundamental than the above quotation indicates. They have been around since the beginning of time. Since there have been "things," there have been relationships among them. And, since humans have been recording information about things, we have been recording relationships. After all, an atom "contains" one or more subatomic particles—a relationship. A figure in a prehistoric cave painting "is adjacent to" some other figure, likely recording some meaningful relationship in the mind of the ancient artist.

This chapter introduces the concept of relationships and gives a brief history of how they have been understood, viewed, and recorded. This historical overview allows us to briefly review topics germane to a complete understanding of this book: class diagrams, relational databases and SQL, and object databases. The chapter also discusses current problems in modeling relationships and recording, or *implementing*, them in a database management system (DBMS). It ends by giving a glimpse at a possible solution.

1.1 Relationships

Precisely, what is a *relationship*? Informally, it is a correspondence, or *association*, among two or more types of things that is identified by a statement relating such things and that may be constrained by certain rules. For example, the statement "an order is placed by a customer" identifies an association between two types of things, orders and customers. This association may be constrained by a rule stating that an order can be placed by only one customer. A *relationship instance* exists when the statement identifying a relationship is true for some specific things and all rules are satisfied. For example, a relationship instance occurs when "order 612 is placed by customer John Doe" and no one else.

Sometimes, a relationship as defined above is called a *relationship type* or *relationship set*. Moreover, a relationship instance is sometimes called a *relationship occurrence*, a *relationship link*, or simply a *relationship*. Using "relationship" for both a relationship type and relationship instance normally causes no confusion. In this

B.K. Ehlmann, *Object Relationship Notation (ORN) for Database Applications*,
Advances in Database Systems 39, DOI 10.1007/978-0-387-09554-7_1,

book, when confusion could result, I use the terms "relationship type" and "relationship instance" (or "*link*") instead of just "relationship."

The reader may skip this paragraph and its mathematics, but formally, a relationship is defined as an *n-ary relation* on *n* sets of things, and a relationship instance is defined as an *n*-tuple. If Orders is a set of orders and Customers is a set of customers, the "is placed by" relationship between orders and customers is defined as

$$\{ (o, c) \mid (o \in \text{Orders} \wedge c \in \text{Customers}) \wedge (o \text{ "is placed by" } c) \wedge \\ \neg\exists c' (c' \in \text{Customers} \wedge c' \neq c \wedge o \text{ "is placed by" } c') \}$$

which is read as "the set of all 2-tuples, or ordered pairs, (o, c) such that o is in Orders and c is in Customers and o is placed by c and there does not exist another customer c' where c' is in Customers and c' is not c and o is placed by c'. Both the statement identifying the relationship and the rule constraining it are part of the conditional statement defining the set of allowable ordered pairs. Within this set, the relationship instance involving order 612 and customer John Doe is represented as (612, John Doe) where 612 is used to uniquely identify the order with order number 612 and John Doe is used to uniquely identify the customer with name John Doe.

Relationship types can usually be viewed from the perspective of any of the types of things being related. For example, instead of stating that an order "is placed by" a customer, we could state that a customer "places" an order. An instance of this relationship, which is the *inverse relationship* of "is placed by," would be (John Doe, 612). (In this paragraph and hereafter only the words that relate things are quoted in the relationship's identifying sentence.)

Relationship types can exist between things that are the same type. For example, a person "is married to" another person. Some of these relationships are *symmetric,* providing an exception to relationship types having different perspectives. For example, a person "is a colleague of" another person can be stated only one way.

All of the examples of relationships given thus far are *binary relationships*, in that they relate two things. Most relationships are binary; however, as our definitions imply, they can be *n*-ary where $n > 2$. A classic *ternary relationship* is that between parts, vendors, and projects: a part "is supplied" by a vendor to a project. An instance of this relationship might be (Type C Gear, ACME Supply, Alpha Project).

Many different relationships often exist among the same types of things. There is, however, usually no interest in recording many of these. For example, the statements that a customer "would like to place" an order and that a customer "would never place" a particular order identify relationships, but an item purchasing system would be unlikely to record them.

1.2 A Brief History of Relationships

We can discover more about relationships by retracing the history of how our knowledge of relationships has evolved; how relationships have been represented, or modeled; and how they have been recorded, or implemented. By "implemented" I mean how they have been stored, used for reference, and kept up-to-date. As we progress through the various data models and data management technologies, which mark the stages of relationship evolution and have largely been defined by them, I offer my perspective on the advancements made and the problems yet to be solved. We begin our review by skipping ahead in time from depictions on cave walls to paper-based systems.

1.2.1 The paper-based system

Before computer file-based systems were invented to record and process data, there were *paper-based systems*. These systems consisted of forms, documents, folders, shelves, and file cabinets. In paper-based systems, there was not much talk of relationships in an abstract sense; instead, they only existed in reality and were recorded by suitable groupings and lists within forms or other documents.

For example, in a paper-based, item purchasing system, the "is placed by" relationship between orders and customers was recorded only by the existence of order forms, each containing entries to record information about one order, as shown in Fig. 1.1. Since there was space to record information about only one customer per form, the rule was enforced that an order "is placed by" only one customer. Of course, the same customer could appear on multiple order forms, such as John Doe as seen in Fig. 1.1. This also means that a customer can place many orders.

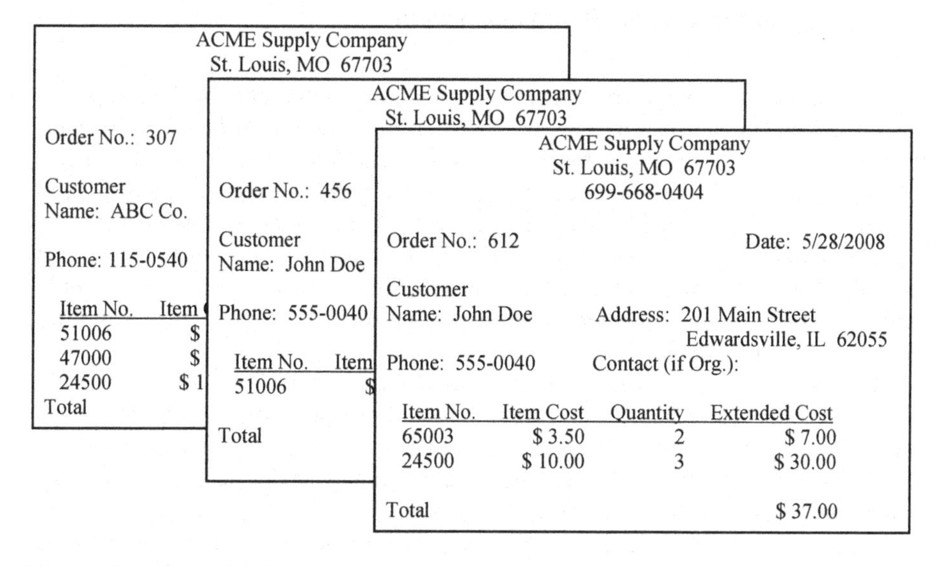

Fig. 1.1 Order forms showing the relationship between orders and customers

The order forms in Fig. 1.1 record more relationships than just the "is placed by" relationship. An order "is placed for" one or more items, as indicated by the list of line items provided on the form, each recording one item. The inverse of this relationship is that an item "is ordered by" an order; actually, it could be ordered by more than one order, such as item number 51006. Some less obvious relationships recorded by the order forms are that a customer "could be" a person and a customer "could be" an organization; or inversely, a person "is a" customer and an organization "is a" customer when placing an order. All of these relationships are addressed later in this chapter in more detail.

Often relationships are recorded by not giving all of the information about the things that are related. Instead, some unique bit of information is given about one or both of the things to allow more information, which is recorded elsewhere, to be referenced and accessed when needed. In our paper-based, item purchasing system, an item catalog likely exists, giving the item number, description, and cost for each item. The Item No. given on an order form references the catalog and can be used to access the item's description in the catalog when needed.

Obviously, in the paper-based system this type of referencing and all of the work needed to keep recorded relationships up-to-date were done by humans in a variety of ways. Relationships were updated by manually creating and deleting forms and documents and making revisions, when possible, to existing ones. The burden of this effort led first to mechanized, punch card systems and then to computerized, file-based systems. Since the former were much like the latter in terms of recording relationships, we skip ahead to the latter.

1.2.2 The computerized file-based system

With the advent of computers in the 1940s and continuing into the 1960s, paper-based (and punch card-based) systems for recording information began to be computerized. A *file-based system* for data recording and processing consisted of a collection of files and programs. Each program needed to have knowledge of, i.e., to declare, the structure of the files that it processed. There was no central control.

In a file-based system, the "system" had no knowledge of the relationships that existed among the data within one file or across files. This knowledge existed only in the minds of the programmers who created and processed the files and was reflected in their program code.

The programs in a file-based system created files with a variety of structures in order to record relationships. Sometimes, physical groupings (i.e., group items) and lists (i.e., arrays) were created that were similar to those used in paper-based systems. Sometimes relationships were recorded in a file by having records of one type sequentially follow each record of a different type.

An example can illustrate some of these possibilities. Fig. 1.2 shows two files that are maintained by a file-based, item purchasing system developed to support or possibly replace the order forms given in Fig. 1.1.

Order File

orderNo	custNo	sub-Rec-Type	date		total	
			lineItem			
			lin	itemNo	itemCost	qty
307	022	0	03/17/2008		636.50	
307	022	1	1	51006	20.95	20
307	022	1	2	47000	6.75	10
307	022	1	3	24500	10.00	15
456	015	0	03/18/2008		125.70	
456	015	1	1	51006	20.95	6
570	005	0	04/02/2008		120.00	
570	005	1	1	24500	10.00	12
612	015	0	05/28/2008		37.00	
612	015	1	1	65003	3.50	2
612	015	1	2	24500	10.00	3

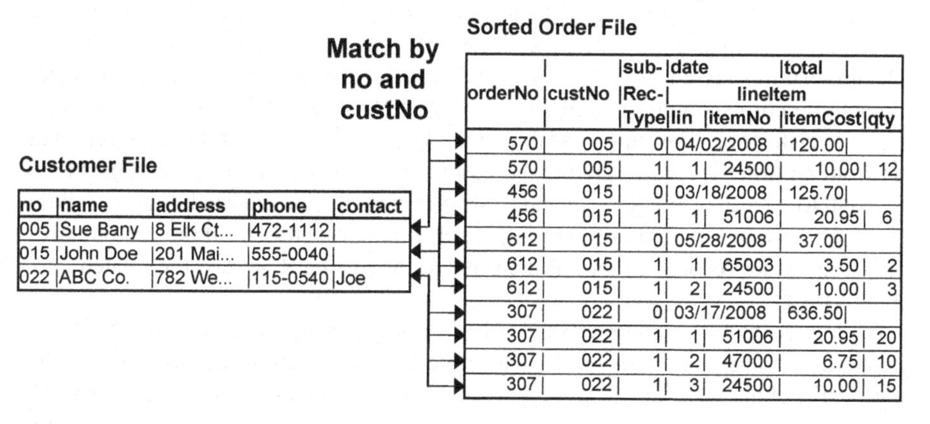

Sorted Order File

orderNo	custNo	sub-Rec-Type	date		total	
			lineItem			
			lin	itemNo	itemCost	qty
570	005	0	04/02/2008		120.00	
570	005	1	1	24500	10.00	12
456	015	0	03/18/2008		125.70	
456	015	1	1	51006	20.95	6
612	015	0	05/28/2008		37.00	
612	015	1	1	65003	3.50	2
612	015	1	2	24500	10.00	3
307	022	0	03/17/2008		636.50	
307	022	1	1	51006	20.95	20
307	022	1	2	47000	6.75	10
307	022	1	3	24500	10.00	15

Customer File

no	name	address	phone	contact
005	Sue Bany	8 Elk Ct...	472-1112	
015	John Doe	201 Mai...	555-0040	
022	ABC Co.	782 We...	115-0540	Joe

Fig. 1.2 Relationships between customers and orders and between orders and line items in a file-based system

An Order file is a sequential file, ordered by two fields within each record, an orderNo and a subRecType. The subRecType indicates whether the remaining fields in a record pertain to the order in general, value 0, or to a line item, value 1. lineItem is a *group item* that contains four fields. lin is a line item number that sequentially numbers the items ordered, itemNo is a unique item number, itemCost is the unit cost of the item, and qty is the quantity of the item being ordered. Records with the same orderNo and subRecType = 1 are ordered by lin.

A Customer file is a sequential file, ordered by no, the customer number. Each record contains the data for one customer. If the customer happens to be an organization, contact is the name of a contact person, else it is blank.

Four relationships are recorded in these two files. The "is placed by" relationship between orders and customers is recorded by having the custNo (customer number) of the customer placing the order recorded in each record of the Order file along with the orderNo (order number). The rule that an order can only be placed by one customer is enforced by the fact that there is only one field provided for a customer number in each order record. The programmer, however, must ensure that the same customer number is placed in all of the records pertaining to an order.

The "is placed for" relationship between orders and items (or inversely, the "is ordered by" relationship between items and orders) is recorded by having records for each ordered item placed sequentially after the "main record" for the order. These records contain the itemNo, itemCost, and qty (quantity) for each item ordered.

The "could be" relationships between customers and organizations and between customers and persons (or inversely, the "is a" relationships) are recorded by simply having a contact field in each Customer file record that means the following: if the customer happens to be an organization, this field contains a name, while if the customer happens to be a person, this field is blank.

In a file-based system, relationships were used for access, using a variety of data processing algorithms that sorted, matched, and/or merged records based on certain fields in the records. For example, to access the name, address, and phone number of the customer placing each order, the Order file was often sorted as shown in Fig. 1.2, and then this sorted file was sequentially accessed along with the Customer file to match up corresponding records based on custNo and no When random access, magnetic disk storage became available in the late 50s, relationship-based access like that from an order record to the related customer record was implemented by 1) using custNo as a disk file key (or index key) to directly (or indirectly via an index) access the customer record or 2) storing the absolute disk address of the customer record in the order record, instead of the custNo, and using it to read the related customer record.

While records were being accessed, they could be updated or deleted. With sequential accessing this often resulted in the creation of new files. Because of the way relationships were physically recorded, deleting a record or records to remove information about one thing from a file often had a side-effect: information about a number of related things was also removed. For example, deleting the Order file records for an order also removed data about the ordered items. If the order deleted was the only order for a particular item (e.g., order 307 and item 47000 in Fig. 1.2), then all information about that item disappeared from the updated file system.

With file-based systems as with paper-based systems, there was not much talk of relationships in an abstract sense; instead, they only existed in reality and perhaps still on forms, and, as we have seen, they were recorded within the computer using a variety of physical organizations within and among files.

1.2.3 Early data models and DBMSs

With the advent of data models and databases in the 1960s, a couple of major advances were made concerning relationships. They began to be viewed abstractly, or logically, and knowledge of them was embedded in the data storage system, now the DBMS. Briefly, a *data model* is a set of concepts for abstractly describing some relevant data, the constraints on this data, and possibly its manipulation. A *database* is a shared collection of logically related, relevant data and a description of this data. A *database management system* (*DBMS*) is a software system that allows and controls the definition and manipulation of a database.

The quotation at the beginning of this chapter implies that relationships played an important role in the development of data models. Indeed, as we shall see, it was largely improvements in our understanding of relationships that ushered in new and improved data models.

The quotation at the beginning of the chapter also implies that relationships are what distinguishes databases from file systems. What is meant here is that databases have a knowledge of relationships while file systems do not. That is, the metadata (i.e., data about the data) that is stored in a database records information about the relationships recorded in the database so that the DBMS can support their use in referencing and update operations. A file system, on the other hand, only records information about the types of things, i.e., the types of *entities*, that are recorded in the files, including for each entity type information about the *attributes* that are recorded. This information is stored in the form of the record and data field definitions given in the file description of each file. The knowledge of relationships is recorded only in the minds of the file system developers and is reflected in the physical file structures they create and the applications code they write to process these structures.

The DBMS—by storing knowledge about relationships, by providing a logical view of them, and by providing central control over the files comprising the database—significantly increased the importance of recorded relationships and improved their handling. Information that before was stored separately in different file-based systems and thus was often incompatible, redundant, and inconsistent could now be integrated into a single database by means of relationships. Also, the operations required to access records via relationships and update relationships could now be done in a uniform manner. These operations were based on having the database designer, application programmer, and user logically view the data (and relationships) in the database in a more abstract manner based on a specific data model.

The earliest DBMSs appeared in the 1960's. IBM's Information Management System (IMS) was based on the hierarchical data model. General Electric's Integrated Data Store (IDS), one of the oldest DBMSs, and Computer Associate's Integrated Database Management System (IDMS), one of the most popular along with IMS, were based on the network data model. The description given of these models and databases draws from a more detailed description given in Ricardo (2004).

1.2.3.1 The hierarchical model and DBMS

The *hierarchical model* uses a tree as its basic logical structure. The hierarchical model for a database is given by a *tree structure diagram*. An example of this diagram, and a possible hierarchical model for our item purchasing system, is shown in Fig. 1.3 (a). A *hierarchical database* consists of a collection of instances of the *type of tree* defined in the tree structure diagram. Each node, or *segment*, of this tree represents a particular type of entity and contains one or more fields representing its attributes. A *parent-child relationship* exists between the entity type represented in the tree by a parent segment, e.g., Customer in Fig. 1.3 (a), and that represented by a child segment, e.g., Order in Fig. 1.3 (a), wherein any instance of a child segment is dependent on an instance of its parent segment and access to the child segments are only provided through the parent segment.

Fig. 1.4 (a) shows one instance of a tree of the type given in Fig. 1.3 (a) within the database. Since Customer is the root segment, the tree instance given is that for customer John Doe. The data given in Fig. 1.2 requires three tree instances, one for each customer. A specific child segment, which always occurs just once in a tree structure diagram, can occur zero, one, or more times in an instance of a tree.

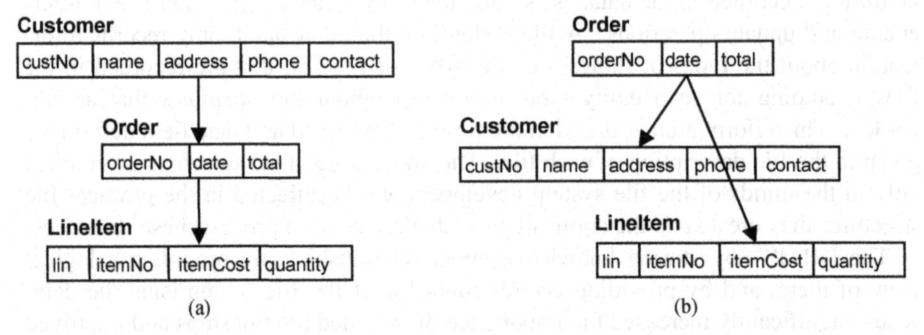

Fig. 1.3 Hierarchical model of item purchasing system, (a) organized by Customer and (b) organized by Order

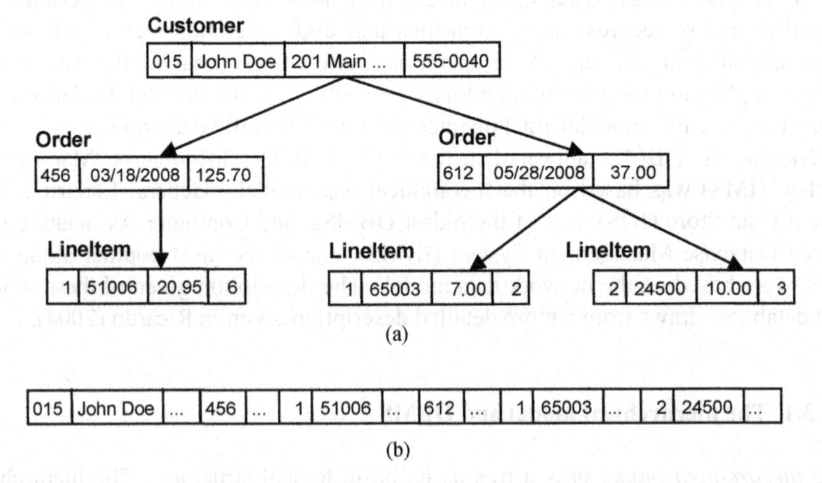

Fig. 1.4 Instance of the Customer tree given in Fig. 1.3 (a), (a) as viewed logically and (b) as stored physically

In the hierarchical database modeled in Fig. 1.3 (a), the "is placed by" relationship between orders and customers is recorded by making an Order segment a child of the Customer segment, thus providing access from a customer to all related orders, i.e., from the *one-side* of the relationship to the *many-side*. The "is placed for" relationship between orders and items is recorded by making a LineItem segment a child of the Order segment, thus providing access from an order to all related items, again from the one-side of the relationship to the many-side.

In the database a tree instance may be stored as one record, where the segments of the tree are stored adjacent to each other in preorder fashion. This storage method is indicated in Fig. 1.4 (b) for the tree instance in Fig. 1.4 (a). Here, the Customer segment is followed by its first Order segment, which is followed by all of its LineItem segments, which is followed by the second Order segment, which is followed by all of its LineItem segments, etc.

In a hierarchical database, searching the database for particular entities involves searching for tree instances and then, if needed, navigating through trees, i.e., doing tree searches. The hierarchical model given in Fig. 1.3 (a) facilitates organizing a file of tree instances so that a Customer segment could be efficiently retrieved given a custNo. (For example, an Index Sequential Access Method (ISAM) file organization with key custNo could be used). This model, however, would be undesirable if efficient retrieval of orders given an orderNo was required. This is because the custNo may not be known and each tree instance, i.e., each file record, may contain many order numbers. A better model for locating orders is given in Fig. 1.3 (b).

This model, however, is problematic in that customer information is redundantly stored. Here, each Order segment has just one Customer segment, but because many orders can have the same customer, many tree instances can contain the same Customer segment.

A major problem with the hierarchical model was that the database designer was forced to record relationships so that one type of entity was the parent, and dominant entity, and the other type was the child, and dependent entity. While this may be natural for some relationships, such as that between an order and its line items, it is unnatural for many others, such as that between orders and customers. The "forced dependency" between related entity types made one type of entity more difficult to reference and access (since access was always through the other type of entity), forced an entity of the type made child to always be connected to an entity of the type made parent, and forced the implicit deletion of a child entity whenever its parent entity was deleted. This implicit deletion can be seen as a forced "referential action." (Similar actions involving relationships are discussed extensively throughout this book.) In addition, if the database designer made the entity type on the many-side of a relationship the parent for better accessibility, like Order in "is placed by," substantial redundancy resulted making updating and the maintenance of consistency very problematic.

Another problem with the hierarchical model was that a segment could have only one parent. Suppose we wanted to computerize the item catalog in our item purchasing system. That is, we would now like to record information about the items that can be ordered and record an "is ordered by" relationship between items and (not orders as before, but more specifically) line items. In Figure 1.3, we would like to add another segment, an Item segment, to either of our models and make LineItem a child of this segment. Unfortunately, this is not possible since LineItem already has Order as its parent segment.

The problems discussed above were addressed to some degree in IMS with *logical databases* and *pointers,* but at the cost of added complexity. These problems did not exist with the network model.

1.2.3.2 The network model and DBMS

The *network model* uses a graph as its basic logical structure. It is also called the *DBTG model* based on the standard developed for the model by the Database Task Group (DBTG) of the Conference on Data Systems Languages (CODASYL). The network model for a database is represented by a *network data structure diagram.* An example of this diagram, and a network model for our item purchasing system, is shown in Fig. 1.5. A *network database* consists of a collection of records of the type defined in the network structure diagram linked together by pointers as also defined in the diagram. Each node, or *record type*, in a network data structure diagram, such as Customer in Fig. 1.5, represents a particular type of entity and contains one or more fields representing its attributes. Each directed edge, such as Cust-Ord, represents an *owner-dependent relationship* between the entity type represented by the record type at the base of the arrow, the owner record type (Customer), and that represented by the record type being pointed to, the dependent record type (Order), wherein any instance of a dependent record is dependent on an instance of its owner record. A single owner record type together with one or more (normally one) dependent record types is called a *set type.* Fig. 1.5 shows three set types: Cust-Ord, Ord-LI, and Item-LI. In a set type, access to the dependent records are provided through the owner record and access to the owner record is provided from the dependent record.

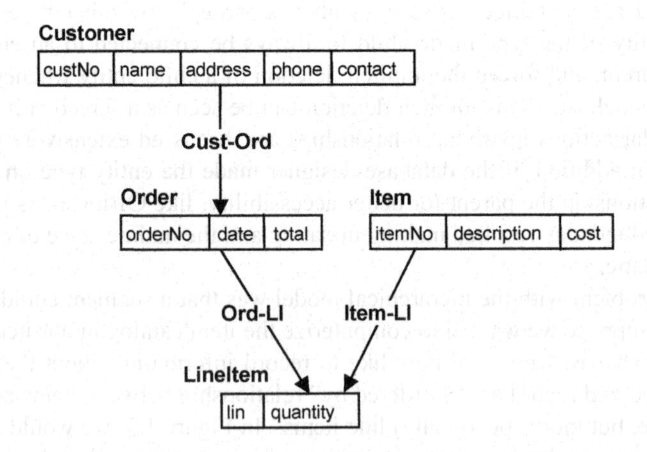

Fig. 1.5 Network model of item purchasing database

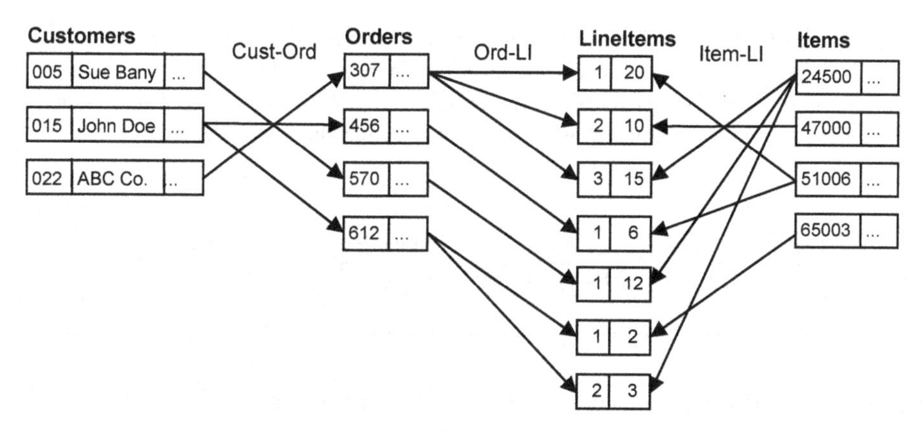

Fig. 1.6 Instance of a database for network model given in Fig. 1.5

Fig. 1.6 shows an instance of a network database, our item purchasing database, consistent with that modeled by Fig. 1.5. Any specific node, which always occurs just once in a network data structure diagram, can occur zero, one, or more times in a database instance.

In the network database modeled in Fig. 1.5, the "is placed by" relationship between orders and customers is recorded by the Cust-Ord set type. This set type, with the Customer record type as owner and the Order record type as dependent, provides easy access from a customer to all related orders, i.e., from the one-side of the relationship to the many-side.

The "is placed for" relationship between orders and items is actually recorded by two set types, Ord-LI and Item-LI. This is because this relationship is actually a bit more complicated than first described. It can actually be viewed as two relationships: an order "contains" one or more line items and an item "is ordered by" one or more line items. (The latter relationship was revealed in the previous section when we attempted to relate an item from the item catalog to a line item.) The Ord-LI and Item-LI set types and set types in general should become more clear when we look at actual instances of the sets defined by these types.

An instance of a set is a collection of records having one owner record and any number of related dependent records, called *members*. In Fig. 1.6, the Order record for order 307, along with its dependent LineItem records with lin values 1, 2, and 3, form an instance of the set Ord-LI. In total, there are four set instances for Ord-LI, one for each Order record. In the DBTG model, a set type is implemented by a circular linked list for each instance, requiring just one pointer in both the owner and dependent record types. (A dependent record can be a member of at most one instance of a given set type.) The implementation of the Ord-LI set instance for order 307 is shown in Fig. 1.7. To access the items for which an order is placed, we access all member LineItem records from an owner Order record via Ord-LI by traversing its linked list. For each LineItem record we come to, we access the owner Item record via Item-LI by traversing its linked list until we come to the owner. Note the amount of navigation! (The DBTG model provides options to add to each member record a direct pointer to its owner or to make the circular linked list a doubly linked list, but these options result in a big increase in the number of required pointers. The model also provides an option to have each owner record point to each member record via a dynamic array of pointers stored in the owner record.)

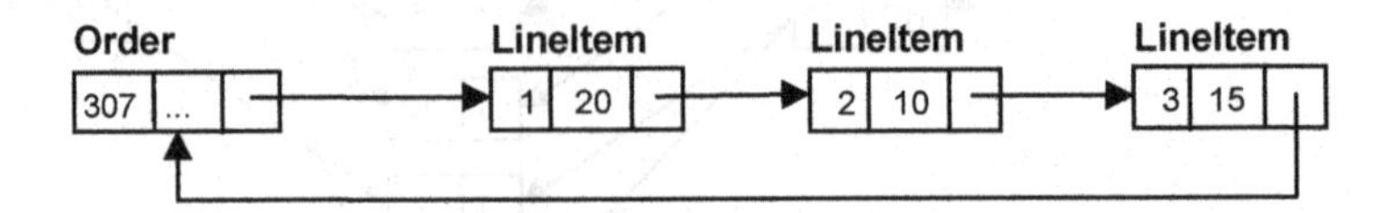

Fig. 1.7 An instance of the Ord-LI set for order 307 implemented as linked list

With the network model, the database designer was still forced to record relationships so that one type of entity is the owner and the other type is the dependent; however, because of certain aspects of the network model, this does not cause the same problems as those caused by the parent/child representation in the hierarchical model. First, in the network model, a record type can have more than one owner, i.e., "parent." Second, all records of a particular record type are in separate files with pointers connecting them, so that access to the records of a particular type, e.g., Order, can be direct and so that there is no structurally imposed rule that each entity of the type made the dependent must be connected to an entity of the type made the owner. And finally, deletion of an owner record type does not necessarily require deletion of all of its connected dependent records.

This is because of the unique referential actions that the model makes available to the database designer, actions that were not forced by the structures used to represent relationships. A *referential action* is a system action or non-action that occurs when an update operation is done that impacts a relationship. This action or non-action may be defined by the model or the DBMS or both. It can be automatic, or it can be optionally specified by the database designer. Referential actions enforce rules that maintain the integrity of the relationship representation and, more importantly, can also serve to define the nature of a relationship.

In the DBTG data definition language (DDL), optional referential actions were defined for each set type and pertained to update operations on owner and member type records. Basically, on insertion of a member record, the possible actions were 1) to do nothing, allowing the programmer to manually CONNECT or not connect the record to an owner record, or 2) to automatically connect the record to an owner record based on a given selection method. On deletion of an owner record, the possible actions were 1) to automatically delete all member records, 2) to automatically disconnect all member records allowing them to exist independently, or 3) to issue an error if there are any member records. On disconnect of a member record from an owner record the possible actions were 1) to do nothing, allowing the disconnect, or 2) to issue an error. On a reconnect of a member record from one owner record to another, the possible actions were 1) to do nothing, allowing the reconnect, or 2) to issue an error.

In summary, the network model and DBMS that evolved during the 1960's and early 1970's offered improvements over the hierarchical model and DBMS in terms of representing and recording relationships. The model offered a logical structure that was more suitable for representing a wider variety of relationship types without data redundancy, and the DBMS supported an impressive set of referential actions to enforce a number of meaningful relationship rules, i.e., *relationship semantics*.

Nevertheless, both approaches retained some characteristics of file-based systems that were problematic. One problem was that accessing records (or segments in the case of hierarchical databases) using relationships required writing code to navigate from record to record through the implementing structures. The second problem was that this code was often dependent on the indexing structures available, the assumed ordering of records, and how the records had been logically structured. This latter consideration was especially true for a hierarchical database. Any change in available indices, ordering, or logical structure resulted in many program changes. These problems were addressed by the relational model and database.

1.2.4 The relational model and RDBMS

The *relational model* and *relational database* were proposed and refined in the 1970's and 1980's by E.F. Codd (Codd 1970, 1980, 1990). Today, the most widely used DBMS is the relational DBMS (RDBMS). The data model was the first to provide a mathematical foundation for a data model and database. Here, I briefly describe how relationships are viewed and recorded in a relational database assuming the SQL:2008 standard (ANSI 2008) minus its object-oriented features.

The relational model uses an *n*-ary *relation* as its basic logical structure. A model for a database is represented by one or more definitions of such named relations, which can simply be viewed as tables. An example of such definitions, and the relational model for our item purchasing system, is given in Fig. 1.8. Each relation (or table) represents a particular type of entity and contains one or more named *attributes* (or columns). Each attribute can be assigned values from a particular *domain*, i.e., set, of values, one value per *tuple* (or row) within the relation. Domain definitions are not given in Fig. 1.8. One or more attributes whose values uniquely identify each tuple are defined as the *primary key* of the relation. Primary key attributes are underlined in Fig. 1.8. Such attributes cannot contain *null* values, a data integrity rule integral to the relational model.

One or more attributes of a relation can also be defined as a *foreign key*. A relation can have more than one foreign key. A relationship exists between an entity type represented by a relation that declares a foreign key, i.e., the *referencing relation*, and the entity type represented by the relation that the foreign key references, i.e., the *referenced relation*. An instance of this relationship exists when a foreign key value in a tuple of the referencing relation matches a primary key value in a tuple of the referenced relation. Foreign key values can be null, unless foreign key columns are defined in SQL as NOT NULL or they are also primary key columns.

Null foreign key values allow entities recorded in a referencing relation to not be related to any entity recorded in the referenced relation. Foreign key values must either match a primary key in the referenced relation or be null, another integral rule of the relational model that enforces what is called *referential integrity*.

Customer(no, name, address, phone, contact);
Item(no, description, cost);
Order(no, date, total, custNo),
FOREIGN KEY custNo REFERENCES Customer;
LineItem(orderNo, lin, itemNo NOT NULL, quantity),
FOREIGN KEY orderNo REFERENCES Order ON DELETE CASCADE,
FOREIGN KEY itemNo REFERENCES Item;

Fig. 1.8 Relational model for the item purchasing system

Customer

no	name	address	phone	contact
005	Sue Bany	8 Elk Ct...	472-1112	
015	John Doe	201 Mai...	555-0040	
022	ABC Co.	782 Wes...	115-0540	Joe

Order

no	date	total	custNo
307	03/17/2008	636.50	022
456	03/18/2008	125.70	015
570	04/02/2008	120.00	005
612	05/28/2008	37.00	015

Item

no	description	cost
24500	Claw Hammer	10.00
47000	25 ft. Hose	6.75
51006	6 ft. Ladder	20.95
65003	Sprayer	3.50

LineItem

orderNo	lin	itemNo	quantity
307	1	51006	20
307	2	47000	10
307	3	24500	15
456	1	51006	6
570	1	24500	12
612	1	65003	2
612	2	24500	3

Fig. 1.9 Instance of a database for relational model given in Fig. 1.8

Fig. 1.9 shows an instance of a relational database, our item purchasing database, consistent with that modeled by Fig. 1.8. In this database, the "is placed by" relationship between orders and customers is recorded by the foreign key custNo in the Order relation. This key and the primary key no in the Customer relation allow the tuple for any order to be matched up, or joined, with the tuple for the customer placing that order. The "contains" relationship between orders and line items is recorded by the foreign key orderNo in LineItem, and the "is ordered by" relationship between items and line items is recorded by the foreign key itemNo in the relation LineItem. These relationship implementations should become more evident after we examine queries that access information using these foreign keys.

A major advance with relational databases was that access to information using relationships was no longer navigational, i.e., record by record; instead, access was specified using a higher-level *query language*. The earliest such languages were relational algebra and relational calculus. SQL, a less mathematical and more English-like language, appeared later. In these languages, database operations would operate on relations rather than records, or tuples. I provide two examples using SQL, the standard DDL and data manipulation language (DML) for relational databases.

Fig. 1.10 (a) shows an SQL query that accesses the order number, date, and total of all orders such that the order "is placed by" customer John Doe. The FROM Order, Customer WHERE Customer.no = Order.custNo specifies that the relations Order and Customer are to be *joined* based on matching the foreign key values of custNo in relation Order with the primary key values of no in Customer. The WHERE ... AND name = 'John Doe' specifies that only the joined tuples containing the value John Doe for the name attribute are to be *selected.* The SELECT Order.no, date, total specifies that only the attributes no from Order, date, and total are to be selected for, or *projected* onto, the resulting relation. The query operates on two relations and returns a third as the result, which is shown in Fig. 1.10 (b).

```
SELECT Order.no, date, total
FROM Order, Customer
WHERE Customer.no = Order.custNo AND
        name = 'John Doe';
```

(a)

no	date	total
456	03/18/2008	125.70
612	05/28/2008	37.00

(b)

Fig. 1.10 (a) SQL query for number, date, and total for each order placed by John Doe and (b) result relation

The query shown in Fig. 1.11 (a) illustrates access using the relational database implementations for all recorded relationships. This query results in a relation containing all of the information recorded in the top order form shown in Fig. 1.1—i.e., it reassembles the order. This query involves three joins of four relations to again produce one result relation, which is shown in Fig. 1.11 (b).

```
SELECT o.no, o.date, c.name, c.address, c.phone, li.lin, li.itemNo,
        li.cost, li.quantity, i.cost * li.quantity AS extendedCost, o.total
FROM Order o, Customer c, LineItem li, Item i
WHERE c.no = o.custNo AND li.orderNo = o.no AND i.no = li.itemNo
        AND o.no = 612
ORDER BY li.lin;
```

(a)

no	date	name	address	phone	lin	itemNo	cost	quantity	extendedCost	total
612	05/28/2008	John Doe	201 Mai...	555-0040	1	65003	3.50	2	7.00	37.00
612	05/28/2008	John Doe	201 Mai...	555-0040	2	24500	10.00	3	30.00	37.00

(b)

Fig. 1.11 (a) SQL query for all information about order 612 and (b) result relation

With the relational model, unlike the network and hierarchical models, the logical structure does not force the database designer to record relationships so that one type of entity is dominant over the other. The structure, however, does force relationships to be recorded so that the referencing relation, the one with the foreign key, represents the many-side of the relationship, if such side exists, and the referenced relation represents the one-side of the relationship, of which at least one must exist. (To model the fact that a many-side does not exist, a foreign key is defined as UNIQUE.) Furthermore, referential integrity forces something to be done when update operations lead to unmatched foreign key values in the referencing relation. As in the network model, what is done is determined by referential actions.

These actions can be defined by the database designer for each foreign key. Again, as with network databases, the choice of actions more precisely defines the nature, or semantics, of a relationship. Basically, on deletion of a tuple in the referenced relation (see ON DELETE clause in Fig. 1.8) possible actions are 1) to do nothing (deferring to a referential integrity check that will be done on transaction commit), 2) to issue an error if any foreign key values in the referencing relation are unmatched, 3) to set any foreign key values now unmatched in the referencing relation to null, or 4) to CASCADE deletion to all tuples in the referencing relation having now unmatched foreign key values. On update of the primary key value in the referenced relation similar actions are possible, except that instead of deleting all tuples in the referencing relation having now unmatched foreign key values, the new primary key value is "cascaded" to these tuples replacing the unmatched values.

The relational model and RDBMS provides a number of advantages in dealing with relationships over previous models and DBMSs. The logical structure provided, including the representation of relationships, is more simple than that of previous models. The operations provided and the query languages based on these operations allow data to be accessed using relationships and manipulated at a higher level of abstraction and in a manner totally independent of how the data is physically stored—i.e., its file organization, indexing, and ordering.

Still many believe the logical structure provided by the relational model is too simple. That is, representing information by placing data into a set of relations (or simple tables) is too constraining, too unnatural, and fails to capture the true semantics, or meaning, of the data. After all, *information* is often distinctly defined as data plus meaning (although I have been using "information" and "data" interchangeably). What is lacking in the relational model is the representation of meaning. To a degree, I believe this is true regarding relationships.

First, for *complex entities*, i.e., those composed of many parts—such as blueprints, Web pages, or even purchase orders—the relational model presents two modeling choices, both problematic. Since column values must be atomic, the first choice is to place the information about each type of entity into multiple rows of a single table resulting in much data redundancy. For purchase orders, this choice is shown in Fig. 1.12. The problem here is that redundancy means wasted space, update anomalies, and possible data inconsistencies. The recommended choice is to use a process called *normalization* to eliminate redundancies by breaking up this table into a set of simpler tables that "properly distribute" the information about each entity. For purchase orders, this choice is shown by the relations shown in Fig. 1.9. Here the problems are 1) the "main table" for each entity type no longer reflects its true nature, e.g., in Order in Fig. 1.9 any indication of line items has been lost, and 2) multiple joins are now required to "reassemble" the entity whenever it is needed for processing or display, e.g., the query for reassembling an order in Fig. 1.11 (a).

Order

no	date	name	address	phone	lin	itemNo	cost	quantity	total
307	03/17/2008	ABC Co.	782 Wes...	115-0540	1	51006	20.95	20	636.50
307	03/17/2008	ABC Co.	782 Wes...	115-0540	2	47000	6.75	10	636.50
307	03/17/2008	ABC Co.	782 Wes...	115-0540	3	24500	10.00	15	636.50
456	03/18/2008	John Doe	201 Mai...	555-0040	1	51006	20.95	6	125.70
570	04/02/2008	Sue Bany	8 Elk Ct...	472-1112	1	24500	10.00	12	120.00
612	05/28/2008	John Doe	201 Mai...	555-0040	1	65003	3.50	2	37.00
612	05/28/2008	John Doe	201 Mai...	555-0040	2	24500	10.00	3	37.00

Fig. 1.12 Unnormalized relation representing an Order.

Second, for relationships, the relational model provides a representation that is, in a sense, "second class" to that of entities. As stated earlier, the model was the first to have a mathematical foundation. The focus, however, was on formally defining entities and the relationships between entity attributes, not on the relationships between entities. Simply representing a relationship as a foreign key having values that may be null or may reference another tuple in a referenced relation disguises and fails to record the true nature and meaning of the relationship. This problem was addressed by the entity-relationship model.

In fairness to the relational model, some relationship semantics can be provided by adding constraints to a database model using a constraint language like SQL. But in practice, the constraints needed are often difficult to formulate and the constraint language as implemented is often not powerful enough to support them. These problems are discussed further in the last section of this chapter.

1.2.5 The entity-relationship model and database

As the name implies, the *entity-relationship (ER)* model finally elevated the status of relationships to that of entities. The model and database were proposed by Peter P. Chen in Chen (1976), a paper that has been one of the most cited in computer science literature. The model presented a higher-level view of information than that presented by previous models—a *conceptual view*, or *conceptual model*, from which previous logical models (and even future ones) could be derived. As such, the ER model or an enhanced version of it is the most popular model used today in conceptually modeling information and databases. In fact, creating an ER model first and then mapping it to a relational model has replaced the normalization process, mentioned earlier, as the preferred way to develop a relational database.

An ER model of a database can be depicted by an ER diagram, which contributed to the popularity of the model. An example of this diagram, and the ER model for our item purchasing system, is shown in Fig. 1.13. (I use the original form of the diagram rather than one of the many "enhanced" forms later proposed by others.) In the diagram, each *entity set* is represented as a rectangle containing its name, and each *relationship set* is represented as a diamond containing its name. A diamond and the lines connecting one or more entity sets indicate that the relationship identified in the diamond is defined among these sets.

In Fig. 1.13, the "is placed by" relationship between orders and customers is represented by the Cust-Ord relationship set, which is defined between entity sets Customer and Order. The "is placed for" relationship between orders and items is represented by the Ord-Item relationship set.

Like the relational model, the ER model is mathematically based. An entity set contains entities sharing common properties, i.e., relationships and attributes. Information about an entity consists of attribute-value pairs, e.g., name-'John Doe' for a customer, where the values for an attribute are drawn from a particular *value set.* (Some ER diagrams may show the attributes of an entity set as labels within ovals connected by a line to the proper rectangle.) A relationship set is mathematically defined as a relation among *n* entities each taken from an entity set. This definition is compatible with the definition of a relationship given at the beginning of this chapter. Fig. 1.14 (a) shows instances of the entity and relationship sets in a tabular form for the item purchasing database. Here each entity is shown with a unique, conceptual identifier followed by the values of its attributes.

The ER model went further than previous models in representing the semantics of relationships. Relationships can have attributes—e.g., the lin and quantity of Ord-Item, recording for each related order and item the line number in the order of the ordered item and the quantity of the item ordered, respectively. The *role* that an entity plays in a relationship can be identified—e.g., an item in the Item entity set when related to an order plays the role of an orderedItem, as is indicated in Fig. 1.13. Also, *n*-ary relationships with $n \geq 3$ can be represented just as easily as binary relationships ($n = 2$) by simply connecting a diamond with three or more rectangles (and expanding relationship sets to include *n*-tuples where $n \geq 3$).

In addition, the *mapping* of a binary relationship can be expressed as 1:1, 1:N, or M:N, meaning a *one-to-one*, *one-to-many*, or *many-to-many* relationship, respectively. This important characterization of a relationship type has only been alluded to in our discussions of previous models since there was no support for such. In fact, the logical structures provided were incompatible with the many-to-many relationship type, which had to be treated as two one-to-many relationships. Now we know that the "is placed for" relationship between orders and items is actually a many-to-many relationship having attributes! Also, the logical structures of previous models had no inherent, or at least obvious, means to constrain the one-to-one type.

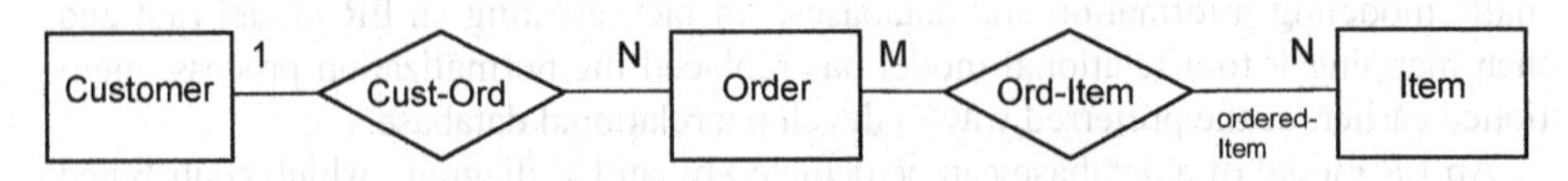

Fig. 1.13 Entity-relationship model for the item purchasing system

Customer set

	no	name	address	phone	contact
c1	005	Sue Bany	8 Elk Ct...	472-1112	
c2	015	John Doe	201 Mai...	555-0040	
c3	022	ABC Co.	782 Wes...	115-0540	Joe

Cust-Ord set

customer	order
c3	o1
c2	o2
c1	o3
c2	o4

Item set

	no	description	cost
i1	24500	Claw Hammer	10.00
i2	47000	25 ft. Hose	6.75
i3	51006	6 ft. Ladder	20.95
i4	65003	Sprayer	3.50

Ord-Item set

order	orderedItem	lin	quantity
o1	i3	1	20
o1	i2	2	10
o1	i1	3	15
o2	i3	1	6
o3	i1	1	12
o4	i4	1	2
o4	i1	1	3

Order set

	no	date	total
o1	307	03/17/2008	636.50
o2	456	03/18/2008	125.70
o3	570	04/02/2008	120.00
o4	612	05/28/2008	37.00

(a)

Entity Relations

Customer

no	name	address	phone	contact
005	Sue Bany	8 Elk Ct...	472-1112	
015	John Doe	201 Mai...	555-0040	
022	ABC Co.	782 Wes...	115-0540	Joe

Item

no	description	cost
24500	Claw Hammer	10.00
47000	25 ft. Hose	6.75
51006	6 ft. Ladder	20.95
65003	Sprayer	3.50

Order

no	date	total
307	03/17/2008	636.50
456	03/18/2008	125.70
570	04/02/2008	120.00
612	05/28/2008	37.00

Relationship Relations

Cust-Ord

custNo	orderNo
022	307
015	456
005	570
015	612

Ord-Item

orderNo	orderedItem	lin	quantity
307	51006	1	20
307	47000	2	10
307	24500	3	15
456	51006	1	6
570	24500	1	12
612	65003	1	2
612	24500	1	3

(b)

Fig. 1.14 Instance of a database for model given in Fig. 1.13 where entities and relationships are viewed (a) conceptually as sets in a tabular form and (b) as logical structures using primary keys

Furthermore, the ER diagram permits the N and M in a mapping to be replaced by a precise upper limit on the number of entities that can participate in relationship instances for each entity in the related entity set (or for each unique set of entities in the related entity sets for *n*-nary relationships, $n \geq 3$). This upper limit is sometimes called a *cardinality*. The ER diagram also permits a lower limit to be given for participation. For example, for Cust-Ord in Fig. 1.13, if the 1 was replaced by 0..1 and the N by 2..15, then we would have a 0..1:2..15 mapping and an order would not have to be placed by a customer and each customer would have to place between two and fifteen orders, inclusive. These are strange semantics from this relationship but, as we shall see, very appropriate semantics for another relationship.

Fig. 1.14 (b) shows a logical structure for our item purchasing system where conceptual identifiers have been replaced by primary keys and entity sets and relationship sets have been replaced by entity relations and relationship relations, respectively. Such an "information structure" was recommended by Chen (1976). Note that if we join the Order entity relation to the Cust-Ord relationship relation based on no and custNo, we get something that looks a lot like a relational database (see Fig. 1.9), which partly explains why no DBMS was ever developed based on the ER model (to my knowledge). Also, the query languages proposed for the ER model were very similar to the relational database query languages. And furthermore, the ER model could be mapped to the hierarchical, network, and relational models, all of which already had DBMS support.

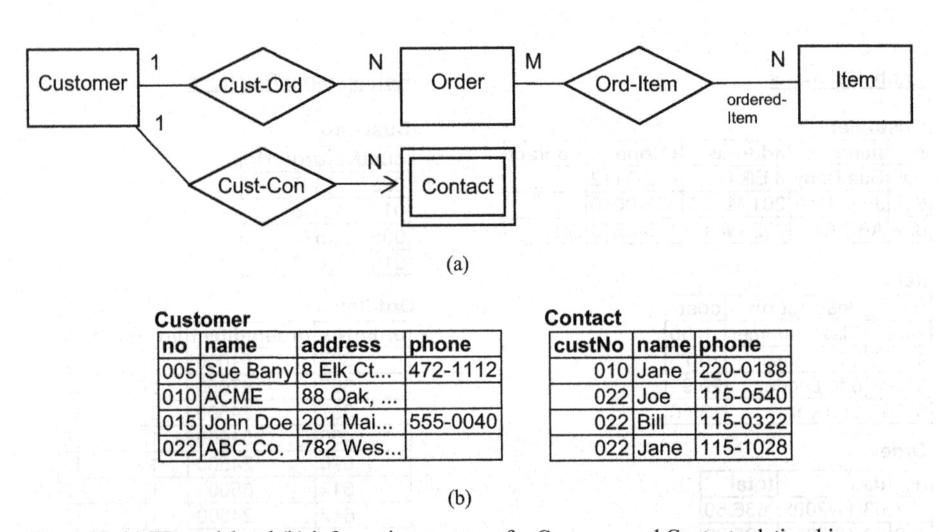

Customer

no	name	address	phone
005	Sue Bany	8 Elk Ct...	472-1112
010	ACME	88 Oak, ...	
015	John Doe	201 Mai...	555-0040
022	ABC Co.	782 Wes...	

Contact

custNo	name	phone
010	Jane	220-0188
022	Joe	115-0540
022	Bill	115-0322
022	Jane	115-1028

(b)

Fig. 1.15 (a) ER model and (b) information structure for Customer and Contact relationship

Before discussing the shortcomings of the ER model, I discuss one more relationship semantic that can be represented. In converting a model's conceptual view of information, like that shown in Fig. 1.14 (a), to the information structure view, like that shown in Fig. 1.14 (b), we may discover that some entity sets have no primary keys; instead, the entities of the set depend on their relationship to another entity for their identity and existence. For example, for an organization that is a customer, we may wish to record multiple contacts each having a different phone number. Fig. 1.15 (a) gives the ER diagram for our item purchasing system updated with a new entity set Contact and relationship set Cust-Con, which represents the fact that a customer "may have" many contacts. A Contact entity is a *weak entity* in this relationship, as denoted by the double line border. Fig. 1.15 (b) gives the new entity relations for Customer and Contact. (I have added one more customer, ACME). The primary key for Contact is custNo and name. The existence of a contact is dependent on its Cust-Con relationship, as denoted by the arrow in Fig. 1.15 (b), so that if a customer is deleted, all related contacts must be deleted.

Despite the many advances made on the relationship front by the ER model, in some respects it fell short. The model as defined in (Chen 1976) provided that "entity sets may not be mutually disjoint," e.g., the set of organizations may belong to the set of customers, yet it did not provide any special representations for such overlapping sets or elaborate on any special impact. This oversight would be addressed later by the semantic and object models.

Also, the Chen (1976) paper in specifically addressing the semantics and rules for data manipulation of entity and relationship sets states:

> It is always a difficult problem to maintain data consistency following insertion, deletion, and updating of data in the database. One of the major reasons is that the semantics and consequences of insertion, deletion, and updating operations usually are not clearly defined; thus it is difficult to find a set of rules which can enforce data consistency.
>
> (Chen 1976, p. 9)

The paper then proceeds to discuss some semantics and rules for data manipulation, including for some operations the "consequences" that "can be performed by the system" to ensure "data consistency." These "consequences," however, which are in fact referential actions, are incomplete and do not enforce the relationship mappings that can be given in an ER diagram. This problem will be addressed by a notation, which is a central topic of this book.

1.2.6 Semantic models, the object model, and the ODBMS

The purpose of *semantic models* was to allow more of the meaning, or semantics, of the data to be represented in the model and ultimately recorded in the database. "More" meant more than that allowed by the relational model, which was considered "semantically impoverished." Even Codd seemed to agree as he proposed adding more semantics to his model in Codd (1979).

Beginning in the late 1970's and continuing into the 1980's there were many semantic models proposed. The ER model could be considered the first and was extended multiple times to make it "semantically richer." Other semantic models proposed were the functional model (Shipman 1981), the semantic data model (SDM) (Hammer and McLeod 1981), and the model for the General Entity Manipulator (GEM) language (Zaniolo 1983). None of these semantic models gained widespread use as the supported model for a DBMS; however, the extended ER models became the basis for the class diagram in the Unified Modeling Language (UML) (OMG 2005), and the logical structure provided by semantic models like SDM became the static structure for the object model, which was supported by multiple object DBMSs (ODBMSs) (Kim 1990, Zdonik and Maier 1990).

Therefore, we continue our history on relationships by turning our attention to the conceptual model defined by the UML class diagram, the object model, and the ODBMS. Even though the class diagram came later, we explore it first since it reflects the relationship semantics discovered by the work on semantic models.

1.2.6.1 Class diagram

In 1997, the UML class diagram was released as part of UML to standardize the many forms of extended ER diagrams that were currently in use. Fig. 1.16 (a) shows a class diagram for our item purchasing database, accounting for the added need of having multiple contacts for an organization. Like the ER diagram, a rectangle provides the name of an entity set or type, which in object-oriented (OO) terms is called a *class*. A class defines a set of entities, which in OO terms are *objects*, that have similar properties and behavior, i.e., semantics. Objects may have *value-based* or *object-based attributes*. These may be listed within the rectangle as shown in Fig. 1.16 (b), a more detailed class diagram. Unlike an ER diagram, *methods* for the class, which define operations on the class and implement behavior, may also be listed for the class. I do not list any of these since our focus is on relationships.

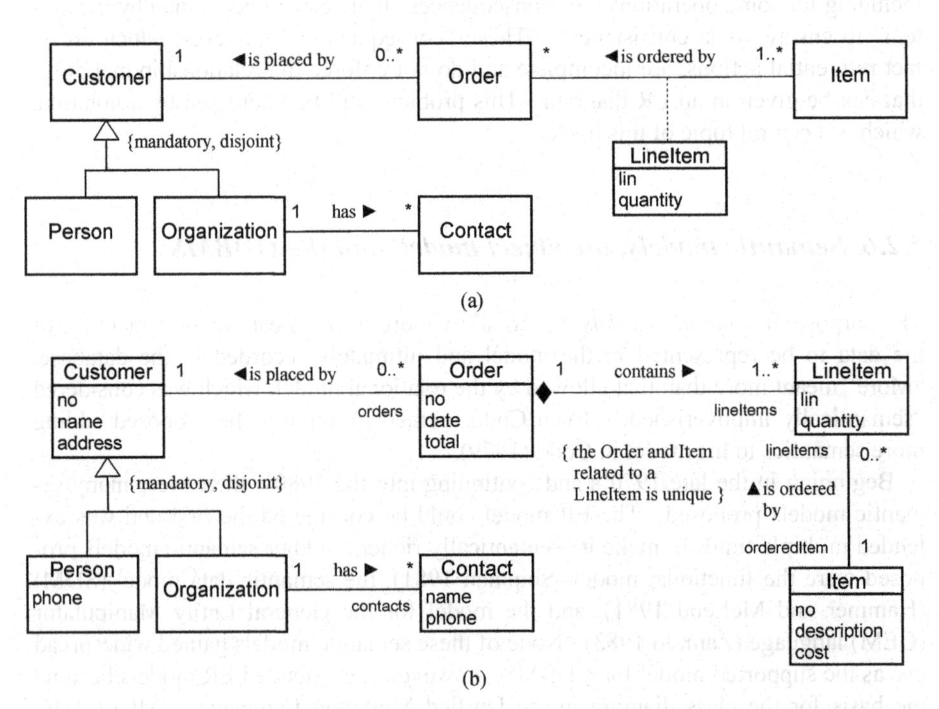

Fig. 1.16 UML class diagrams for the item purchasing system, (a) initial model and (b) revised and more detailed model

Relationships of the type called *associations* in UML are represented much like relationships in an ER diagram, except that the diamond is eliminated and a phrase that names the association, e.g., "is placed by," is put near the middle of the line linking the two related classes. (Ternary and higher degree relationships are depicted much like they are in an ER diagram except that the diamond is smaller and doesn't contain any identifier.) A *readability indicator*, e.g., ◀, can be placed next to an association name to indicate the direction in which the association should be read, e.g., "an Order is placed by 1 Customer."

The "1" is a *multiplicity*. Multiplicities are placed at each *association end* (next to the classes for which they apply) and in ER terms provide the "mapping" for the relationship. A multiplicity for a class A indicates how many objects of type A can be related to one object of the related class (or to a fixed set of $n - 1$ objects each taken from one of the $n - 1$ related classes for an n-ary association where $n \geq 3$). In general, multiplicities can be given as $n..m$, where n is the lower bound and m is the upper bound or *, which implies no limit. A 1 means 1..1 and a * means 0..*.

Like the ER diagram, a role name can be given at an association end, e.g., orderedItem in Fig. 1.16 (b). A role name indicates the role that an object plays when involved in the association. As we shall see, these names can play an important "role" when a class diagram is mapped to an object model.

An important advancement with semantic models was the recognition that an *"is a" relationship* (or inversely, a "may be a" relationship) between two types of things, e.g., an organization "is a" customer, is fundamentally different from other types of relationships (Smith and Smith 1977). Here, the two things being related are basically one and the same. That is, in ER terminology, one entity set is a subset of the other entity set, or in OO terminology, one class is a *subclass* and the other is a *superclass*. Since an object in a subclass is an object in a superclass, it inherits the attributes and relationships (and methods) of the superclass object. Also, because of its "being singled out for special treatment," it may have its own attributes and relationships.

In UML, an "is a" relationship between a subclass, e.g., Organization, and a superclass, e.g., Customer, is denoted as shown in Fig. 1.16. A "white" arrowhead points to the superclass. It is assumed that any attributes listed for a superclass—e.g., no, name, and address for Customer in Fig. 1.16 (b)—and any associations involving the superclass, e.g., "is placed by," are inherited by a subclass, e.g., Person and Organization. Association names, multiplicities, and roles are not applicable to an "is a" relationship. The {mandatory, disjoint} constraint given in Fig. 1.16 indicates that a Customer must be either a Person or Organization and cannot be both. In class diagrams constraints such as these are given in braces.

So, to be precise, we must now distinguish between two types of relationships: "is a" relationships and associations. The "is a" relationship is also called a *generalization/specialization relationship* because a superclass is a generalization of its subclasses and a subclass is a specialization on its superclass(es). An *association* is sometimes called a *structural relationship*—because it is often represented structurally, e.g., using trees, networks, or embedded foreign keys or pointers. An "is a" relationship, on the other hand, is usually represented by some programmer convention within a record or tuple, e.g., a contact attribute is blank if the customer represents a person and non-blank if an organization, or it must be defined to and properly handled by the system as specific type objects are created and manipulated. It is important to note that an association "is a" relationship, but the inverse is not true. I sometimes still refer to an association as a relationship in the remainder of this book.

In Fig. 1.16 (a), an *association class* is given for the many-to-many association "is ordered by." An association class can be given for any association to indicate that the association has some attributes, e.g., lin and quantity for the "is ordered by" association. (Often, the name of the association class is omitted and only association attributes are listed.) While many-to-many associations, including an association class, can also be represented directly in an ER model and database (see Ord-Item in Figs. 1.13 and 1.14), they must be viewed and recorded in other models and databases as two one-to-many associations. These one-to-many associations share an *intersecting relation* in the relational model and a "class that represents the association," or an *associating class*, in the object model.

For this reason, the "is ordered by" association in Fig. 1.16 (a) has been remodeled in Fig. 1.16 (b). Here, the associating class is LineItem. A constraint is given on this class to ensure that our remodeling preserves the semantics of the original association. In UML, an association is a **set** of tuples where each element in a tuple represents an object of one of the related classes, and being a set, it can contain no duplicate tuples. So, while Fig. 1.16 (a) does not allow a particular order and item to be related twice, Fig. 1.16 (b) would (perhaps with different lin or quantity values) if it were not for the given constraint.

In addition to the "is a" relationship, a class diagram provides for the representation of another important kind of relationship type. These are associations that have semantics based on the concept of "whole-part," i.e., the concept that one thing "is a part of" another thing. The diamond in Fig. 1.16 (b), placed at the Order end of the "contains" association, indicates that an order is an *aggregate* and a line item is a *component* in this association. What this means mathematically, according to the UML specifications (OMG 2005), is that the relationship between the aggregate class, here Order, and the component class, here LineItem, is an *asymmetric* and *transitive* relation. (Asymmetric means that if *a* contains *b*, then *b* does not contain *a*. Transitive means that if *a* contains *b* and *b* contains *c*, then *a* contains *c*.) *Aggregation* also means something that is more difficult to define mathematically, simply the sense that a specific component, here a line item, "is a part of" a specific aggregate, here an order. (Chapter 5 contains a more detailed discussion of these concepts.)

Actually, two types of aggregation can be represented in a class diagram. The strong type, called *composition*, is represented by a darkened (black) diamond, as in Fig. 1.16 (b). Here, there is an added semantic, or rule, that a component can only be part of one aggregate (no matter what the type). Obviously, a line item can only be part of one order, so composition applies and an order is a *composite object*. The other type of aggregate is a weaker form called *shared aggregation* and is represented by an unfilled (white) diamond. Here, a component can be part of more than one aggregate, e.g., a type of engine can be part of more than one type of vehicle.

Like the ER diagram, the class diagram provides a conceptual model that can be mapped to other logical models. Today, the generally adopted methodology for database development first builds a conceptual model using a class diagram, then maps this to a logical model, either relational or object, and then "implements" this model using a target DBMS. Next, we review the object model and ODBMS and map the class diagram for our item purchasing system to an object model and database implementation.

1.2.6.2 The object model and ODBMS

Object databases began to appear in the mid 1980s (Copeland and Maier 1984) and many ODBMSs were developed in the late 1980s and early 1990s. Their development was motivated by a need to record and efficiently manipulate complex objects, e.g., blueprints and Web pages, a need not being met by relational databases. Another driving force was the development of OO programming languages. Object databases and ODBMSs combine the richer modeling concepts and constructs of semantic models—like unique, attribute-independent entity identifiers, type hierarchies, inheritance, and single and multiple valued, entity-valued attributes—with OO features like data abstraction and object behavior, i.e., the dynamic aspects of user-defined operations or methods. In 1993, the Object Data Management Group (ODMG), a group supported by ODBMS vendors, released a standard for an Object Model, Object Definition Language (ODL), and Object Query Language (OQL) (Cattel et al. 2000). Most of the discussion below is based on this standard.

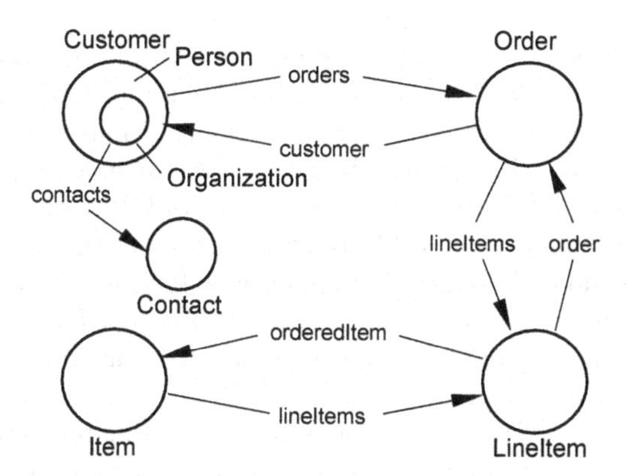

Fig. 1.17 Semantic/object model of the item purchasing system based on class diagram in Fig. 1.16

Fig. 1.17 is a directed graph that shows how the classes and relationships in Fig. 1.16 are represented in an object model. (Actually, this type of diagram was often used for the semantic model.) In the graph, each node represents a class. A directed edge from a class *A* to a class *B* represents an *object-based attribute* within class *A* that records a relationship between *A* and *B* from the perspective of class *A*. For each object of class *A*, the attribute references all related objects of class *B* and thus provides an *access path* from *A* to *B*. If the attribute name is singular, it is a *single-valued attribute*, and if plural, it is a *multi-valued attribute*. Normally, each object-based attribute has an *inverse attribute* that records the relationship from the perspective of the related class and provides the inverse access path from an object *B* to an object *A*. Role names given in a class diagram map to object-based attributes.

For example, in Fig. 1.17 orders is a multi-valued, object-based attribute in class Customer that references all of the orders placed by a particular customer. Its inverse, customer, is a single-valued, object-based attribute in Order that references the customer that placed the order. The multi-valued, object-based attribute contacts in class Organization has no inverse, which means there is no access path from a Contact object to an Organization object. In fact, since there is no inverse, the relationship between organization and contacts may not be implemented as a true object database relationship.

Fig. 1.18 shows a partial Object Database Definition Language (ODDL) specification that corresponds to the model in Fig. 1.17. ODDL is a language that is closely compatible with the ODMG standard ODL. Fig. 1.19 shows an instance of our item purchasing database, consistent with that modeled by Figs. 1.16 and 1.17 and defined by Fig. 1.18. An important object database concept is that each object in the database is assigned a unique, immutable identifier by the system, which can be used to reference the object. This identifier is labeled as id in Fig. 1.19.

Customers, Orders, and Items defined at the end of the specification in Fig. 1.18 are class *extent* objects. Values for these objects are shown at the top of Fig. 1.19. An extent is a *collection* type object that contains references to all objects of a particular class. There are a number of collection subtypes—e.g., Set, List, and Array—some of which allow a specified ordering of objects. e.g., List. In Fig. 1.18, each of the extents is defined to be a Set.

Single-valued, object-based attributes are defined as a reference type, which can reference an object of a particular type. For example, in Fig. 1.18 the attribute customer in class Order can reference a Customer object. Multi-valued, object-based attributes, like extents, are defined as collections. For example, in Fig. 1.18 the attribute orders in class Customer is defined as a Set of Order references. The object-based attribute contacts in class Organization is defined as a List; however, it is not known as a relationship to the ODBMS since no inverse clause is given. Such alternative "relationship implementations" are allowed in an object database because, unlike relational databases, attribute values need not be atomic, and there are a wealth of available attribute types.

Object databases provide two ways to access objects using relationships, one procedural and the other non-procedural. Both ways employ *path expressions*, which use a dot notation introduced by semantic models (Zaniolo 1983). A dot operator allows access from one object to another via a reference. From a relational database perspective, a dot operator provides a "more convenient join." A path expression may contain many dot operations, the last of which may specify access to a value-based attribute. For example, the path expression currentLI.order.customer.name accesses the name of the customer who placed the order containing the line item referenced by the variable currentLI.

```
...
class Customer {
  int no;
  string name;
  string address;
  Set<Order> orders inverse customer;
  ...
};
class Person isa Customer {
  string phone;
  ...
};
class Organization isa Customer {
  List<Contact> contacts;
  ...
};
class Item {
  int no;
  string description;
  float cost;
  Set<LineItem> lineItems
    inverse orderedItem;
  ...
};
class Order {
  int no;
  Date date;
  float total;
  Customer customer inverse orders;
  List<LineItem> lineItems inverse order;
  ...
};
class LineItem {
  int lin;
  int quantity;
  Order order inverse lineItems;
  Item orderedItem inverse lineItems;
  ...
};
class Contact {
  string name;
  string phone;
};
extent Set<Customer> Customers;
extent Set<Order> Orders;
extent Set<Item> Items;
...
```

Fig. 1.18 Partial ODDL specification for the model shown in Fig. 1.17

Customers

{c1, c2, c3, c4}

Orders

{o1, o2, o3, o4}

Items

{i1, i2, i3, i4}

Person Objects

id	no	name	address	phone	orders
c1	005	Sue Bany	8 Elk Ct...	472-1112	{o3}
c3	015	John Doe	201 Mai...	555-0040	{o2, o4}

Organization Objects

id	no	name	address	contacts	orders
c2	010	ACME	88 Oak, ...	t1	{}
c4	022	ABC Co.	782 Wes...	t2, t3, t4	{o1}

Item Objects

id	no	description	cost	lineItems
i1	24500	Claw Hammer	10.00	{l3, l5, l7}
i2	47000	25 ft. Hose	6.75	{l2}
i3	51006	6 ft. Ladder	20.95	{l1, l4}
i4	65003	Sprayer	3.50	{l6}

Contact Objects

id	name	phone
t1	Jane	220-0188
t2	Joe	115-0540

id	name	phone
t3	Bill	115-0322
t4	Jane	115-1028

Order Objects

id	no	date	total	customer	lineItems
o1	307	03/17/2008	636.50	c4	l1, l2, l3
02	456	03/18/2008	125.70	c3	l4
o3	570	04/02/2008	20.00	c1	l5
04	612	05/28/2008	37.00	c3	l6, l7

LineItem Objects

id	order	lin	Ordered item	quantity
l1	o1	1	i3	20
l2	o1	2	i2	10
l3	o1	3	i1	15
l4	o2	1	i3	6
l5	o3	1	i1	12
l6	o4	1	i4	2
l7	o4	2	i1	3

Fig. 1.19 Instance of database consistent with semantic/object models given in Figs. 1.16 and 1.17 and ODDL given in Fig. 1.18

Objects are accessed procedurally using an *application program interface (API)* provided by the ODBMS and embedded in an object-oriented programming language, like C++ or Java. The object navigation required is somewhat reminiscent of record navigation required in a network database. Navigation using single-valued, object-based attributes is made easy using path expressions. Navigation using multi-valued, object-based attributes, however, can be a bit more complicated. An *iterator* object is declared on the collection. Then, a looping construct is employed to sequentially access objects referenced by the collection using methods provided by the iterator. Some collections allow random access by key.

A nonprocedural way to access objects is provided by a query language. I give two examples using the OQL, which is the ODMG standard query language for object databases and is similar to SQL.

```
select struct(no: o.no, date: o.date, total: o.total)
from Orders o
where o.customer.name = "John Doe"
```

Fig. 1.20 OQL query for number, name, and total for each order placed by John Doe

Fig. 1.20 shows an OQL query that accesses the order number, date, and total cost of all orders such that the order "is placed by" customer John Doe. We can compare this query to its relational counterpart, shown in Fig. 1.10 (a). While the data that results is the same as that shown in Fig. 1.10 (b), the type of the result is not the same. Instead of resulting in a relation, this query results in an instance of a structure for each order and returns a collection of these instances of the type:

```
Set<struct(no: int, date: Date, total: float)>
```

```
select struct(no: o.no, date: o.date,
        name: o.customer.name, address: o.customer.address, phone: o.customer.phone,
        lineItems: (select struct(lin: li.lin, itemNo: li.item.no, cost: li.item.cost,
                                quantity: li.item.quantity,
                                extendedCost: li.item.cost * li.item.quantity)
                    from o.lineItems),
        total: o.total)
from Orders o
where o.no = 612
```

Fig. 1.21 Query for all information about order 612

The query shown in Fig. 1.21, illustrates access using the object database implementations for the "is placed by," "contains," and "is ordered by" relationships involving orders, customers, line items, and items. We can compare this query to its relational counterpart, shown in Fig. 1.11 (a). Again, while the data that results is the same as that shown in Fig. 1.11 (b), the type of the result is not the same. Instead of resulting in one relation to represent an order, which "flattens it out" and duplicates data, this query results in an instance of a complex structure that better represents the natural structure of an order, which is a complex object. Specifically, it returns a struct of the following type, involving no data duplication:

```
struct(no: int, date: Date, name: string, address: string, phone: string,
      lineItems: Set<struct(lin: int, no: int, cost: float, quantity: int, extendedCost: float)>,
      float total)
```

It is interesting that even though object databases can better represent complex objects (e.g., the definition of an Order object in Fig. 1.18 clearly includes a list of line items), the individual parts of a complex object, like an order, must still be accessed and "reassembled" from the database, as with relational databases. Path expressions, however, provide a more natural way and usually a more efficient way to access these parts than do joins.

It is also interesting that the data independence achieved with relational databases is compromised with object databases. When relationships are used to access related objects, the order of access on the many side of a relationship is determined by the type of collection defined to implement that side. If program or query code is dependent on a known ordering provided by a collection, e.g., List, and the ordering is later changed, perhaps because of a change in collection type, e.g., to a Set, then this code must be changed.

The ODMG object model enforces referential integrity for relationships by providing some automatic referential actions. Here, "referential integrity for relationships" means no dangling references (i.e., no references to objects that do not exist) in any attribute used to implement a relationship. If an object is deleted, any single-valued attribute that references the object is set to null and any reference to that object in any multi-valued attribute is removed. If a change is made to any attribute that implements a relationship, an appropriate corresponding change is automatically made to its inverse attribute. For example, if currentOrder.customer is set to null, a reference to currentOrder is automatically removed from its inverse attribute currentOrder.customer.orders. Referential actions do not apply to object-based attributes that do not implement a true relationship, e.g., contacts in class Customer.

What is missing from the standard object model and ODBMS are additional, developer-specifiable referential actions. Even the equivalent actions to those available in the ON DELETE clause of the foreign key declaration in SQL are missing. Such actions would, for example, automatically delete all related LineItem objects when an Order object is deleted or cause an exception when an attempt is made to delete a Customer object if related Order objects exist. Apparently and unfortunately, it is assumed that such actions will be implemented by programmers within appropriate class methods.

1.2.7 The object-relational DBMS

Into what is basically a relational DBMS, an *object-relational DBMS* includes some object-oriented features, primarily support for the "is a" relationship and the ability to declare a tuple or attribute to be a type of object. The SQL:2008 standard includes such features (ANSI 2008). This inclusion has resulted in no advances in terms of the understanding, modeling, or recording of relationships. So this section is very short, offering only my perspective.

The object-relational DBMS has resulted in more flexibility in that relationships can now be implemented using a number of techniques—foreign keys, references, and embedded structures within objects. Unfortunately, it has also resulted in more complexity in that all of these techniques must be documented, supported by different data query and update operators, and understood by database developers. In addition, the database developer is now presented with the dilemma of deciding which implementation techniques to use for which relationships, decisions that are often difficult and not made any easier by the DBMS. If decisions must later be changed, database restructuring and query and application program changes are often required.

1.3 Problems in Modeling and Implementing Relationships

As already mentioned in the discussion on the UML class diagram, the current technology for developing databases first builds a conceptual model using a class diagram, then maps this model to a logical model, and then "implements" this model using a target DBMS. The logical model and DBMS are usually relational, object, or object-relational, but are most often relational. This section discusses relationship-related problems that still exist with the current technology—specifically, with data models and DBMSs—that make database development overly difficult and prone to error. We have seen the fundamental role that relationships have played in defining models and DBMSs. Yet, despite this role, their mapping from their modeled representations to implementation is not straightforward and can be quite difficult.

One reason for this stems from the fact that the technology for implementing relationships using a DBMS has not caught up with the technology for modeling them using a class diagram. Since the inception of the ER model (Chen 1976), database developers have been modeling relationships using one paradigm and then have been forced to implement them using a totally different one, that offered by the DBMS. As a result, the one-to-one, one-to-many, and many-to-many notations for relationships—and particularly, their more precise mapping or multiplicity semantics—given in an ER or class diagram must be essentially abandoned when it comes time to define relationships to a DBMS. Consequently, they must be defined indirectly in DDLs or often implemented in code with great difficulty in both relational and object DBMSs (Ehlmann and Yu 2005b).

In an RDBMS, a many-to-many relationship must, of course, always be implemented as two one-to-many relationships, and a one-to-one relationship implemented as a UNIQUE constraint on the implementing foreign key. The implementations of precise multiplicity constraints, however, are more problematic. The only multiplicity constraint that is easy to implement, is a lower bound of 1 on the one side of a relationship. This can be implemented by defining a NOT NULL constraint on the implementing foreign key, provided the referenced relation is the 1 side of the relationship. (This need not always be the case in a one-to-one relationship—e.g., an Employee relation, the referencing relation, contains a foreign key issuedCard that **may** reference a credit card number in the CreditCard relation, the referenced relation, and each credit card must be assigned to 1 employee). Other multiplicity constraints must be implemented by complex constraints and triggers as we shall soon discover by an example.

In an ODBMS, the one side of the relationship is implemented as a reference and the many side as a collection. A many-to-many relationship must be implemented as two one-to-many relationships if it has some attributes. The implementation of precise multiplicity constraints is totally the responsibility of the database developers. The constraints must be enforced by properly coding all class methods that deal with the inserting or deleting of objects or the initial setting or updating of any object-based attribute. The appropriate checks must be made in these methods and the proper exceptions must be raised .

Another significant reason why relationship implementation can be difficult with current technology is a dearth of declarative referential actions available to database developers. Such actions, if specifiable in a class diagram, would allow database designers to better model relationships and, if supported by the DBMS, would allow developers to more easily implement them. As previously discussed:

- The ER diagram (and model) allows no optional referential actions to be specified except those related to weak entity types. When an entity type is made weak, then presumably an entity of this type is automatically deleted on deletion of its related "strong" entity, and on insertion of a weak entity, an error is given if the entity is not linked to an entity of its strong entity type.
- The class diagram, like the ER diagram, allows no optional referential actions to be specified except that some referential actions are assumed when a class is made a component class of a composite class or a subclass of a superclass. In the former case, on deletion of a composite object, any related component object is deleted, and on linking a component object to a second composite object, an exception is raised. In the latter case, on deletion of a subclass object, the superclass object is deleted and vice versa. These "objects," however, are really the same objects, so these actions cannot really be called referential actions. (Stereotypes, e.g., «weak entity», can be used in class diagrams to label classes as weak or strong entity types. This presumably indicates the same referential actions as those for ER diagrams.)
- SQL has ON DELETE and ON UPDATE referential actions for foreign key declarations. These, however, only deal with tuple deletion and primary key update operations on one side of a relationship, i.e., on the referenced relation side. They do not provide **any** actions relevant to tuple deletion or update foreign key operations on the referencing relation. These foreign key operations are important in that they result in the creation and destruction of relationship instances.
- The ODMG object model provides no optional referential actions for associations, period! The "referential actions" for subclass/superclass objects are the same as those inferred by a class diagram.

I use the words "presumably" and "assumed" above in describing the referential actions supported by the ER and class diagrams because both are only models and without corresponding support from a DBMS, the actions are not automatic, but must be manually implemented by the database developer. Productivity improves immensely when the DBMS supports the declarative referential actions of a model.

Today, the class diagram model does not provide referential actions relevant to the enforcement of multiplicities, and DBMSs do not support the enforcement of multiplicities or any relevant referential actions, i.e., "multiplicity actions." Why can't multiplicities be specified in the DDL of a DBMS and be enforced by the DBMS? Why, for example, can't a system action like CASCADE be specified for object deletion when this deletion would violate a lower bound multiplicity? As we shall see in subsequent chapters, such enforcement and related actions can better define the nature, i.e., semantics, of relationships and thus improve the productivity of developing database systems and enhance database integrity.

An example will illustrate the current difficulty of modeling certain relationships in a class diagram and implementing them in a relational database. Fig. 1.22 models part of a company database. Relationships exist wherein a unit within the company "consists of" zero or more units, an employee "works for" a unit, and an employee "belongs to" zero or one carpool. The "consists of" association is an example of an *intra-class association* since it involves just one class. It is also an example of a *recursive association* since a child unit can itself be a parent unt. A unit may consist of units, each of which may consist of units, each of which may consist of units, … .

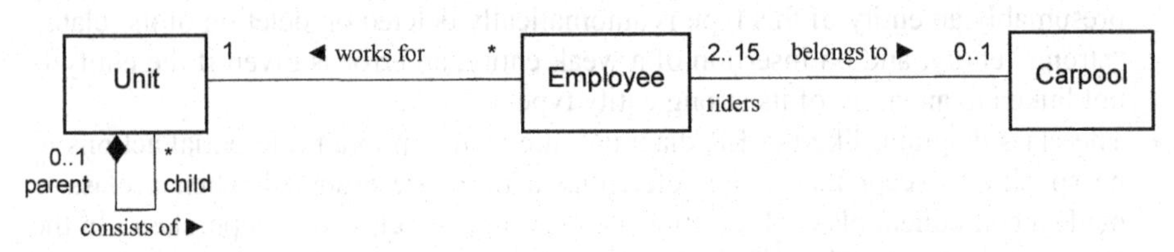

Fig. 1.22 Class diagram for part of a company database

Besides the semantics described in the diagram, here are some additional ones relevant to the relationships that cannot be described in the diagram:

- An attempt to remove a unit that has employees working for it should cause an error since all employees must work for a unit. (This semantic may seem to be implied by the 1 for Unit, but employees could just be removed along with the unit.)
- If a unit is removed, all subordinate, i.e., component, units should be automatically removed unless those units still have employees working for them. Such units may remain independent or later be placed within another unit.
- If a unit is removed from its parent unit, similar semantics should apply. It and all of its subordinate units should be automatically removed unless they have employees working for them.
- A carpool is defined by having at least two riders, so that if the number of riders falls below two, due to employee terminations or employees quiting the carpool, the carpool really no longer exists and so should be automatically removed.

These kinds of semantics are often important in better defining relationships.

Fig. 1.23 shows an **attempted**, standard SQL implementation of the model in Fig. 1.22 and the above relationship semantics. The code specifically written to implement these semantics is given in italics. The code was generated by a modeling tool, which is why primary keys are named pk, instead of more appropriate names, like ssn in table Employee. The reader is not expected to fully understand this code. I give it only so the reader can appreciate the difficulty of writing and testing such code. It involves a complex check constraint and multiple triggers and yet still falls short of adequately implementing the relationship semantics as described above. For those readers having SQL expertise, here is a brief description of its shortcomings.

- The code uses a check constraint with a nested query that references another table to ensure that the multiplicity lower bound of 2 for the "belongs to" relationship is satisfied at transaction commit. The problem is that such nested queries are not allowed in the SQL dialects of most existing RDBMSs. This means triggers must be written to check lower bound multiplicity constraints, but such checks done in triggers cannot be deferred until transaction commit.
- In general, checks for lower bound multiplicity violations and partial rollbacks cannot be done when needed because nested transactions cannot be defined in standard SQL and needed constraint checking cannot be tracked and controlled. Without going into the details, these limitations mean that performing more than one complex object operation, like removing a unit, in a single transaction is problematic and worse that semantics such as those given in the second and third bullets above cannot be implemented in standard SQL, at least not via triggers.
- (As many SQL dialects require referenced identifiers to be predefined, the definition of some foreign keys and constraints must be done by alter statements.)

1.4 Preview of a Solution

I close this first chapter by giving a sneak preview of what lies ahead, namely a solution to the problems just discussed. Fig. 1.24 (a) shows an extended class diagram, which includes notations that provide the desired referential actions that better define the modeled associations. These are the ' binding symbol given at the parent end of the "consists of" association, the ? binding symbol given at the Employee end of the "belongs to" association, and the lack of binding symbols given at all other association ends.

Fig. 1.24 (b) gives an extended SQL that implements all of the desired association semantics. Again, the code specifically written to implement these semantics is given in italics. All that is needed is a notation for each foreign key definition, e.g., ?<2..15-to-0..1> for foreign key fk_Carpool_rO, that directly maps the class diagram representation of an association to its SQL definition, which is also its implementation since the DBMS will now enforce all multiplicities and execute all referential actions. No complex constraints, triggers, or programming is required.

```
-- sO - subject Object  rO - related Object
-- pk - primary key     fk - foreign key
create table Unit (
  pk          char(2),  --or another primary key
  --other attributes
  fk_Parent   char(2) --fk for "consists of"
              references Unit(pk)
                on update cascade
                initially deferred,
  primary key pk
);
create table Carpool (
  pk          char(2),  --or another primary key
  --other attributes
  check( (select count(*)
          from Employee rt
          where rt.fk_Carpool_rO = Carpool.pk)
            >= 2 ) initially deferred,
  primary key pk
);
create table Employee (
  pk          char(2),  --or another primary key
  --other attributes
  fk_Unit_sO  char(2) --fk for "works for"
              not null initially deferred
              references Unit(pk)
                on update cascade
                on delete set null
                initially deferred,
  fk_Carpool_rO  char(2) --fk for "belongs to"
              references Carpool(pk)
                on update cascade
                on delete set null
                initially deferred,
  primary key pk
);
create trigger T_Delete_Unit
  after delete on Unit
  referencing old row as old for each row
  begin atomic
    declare binding_exc condition;
    begin
      declare c cursor for
        select pk
        from Unit rt  -- rt - referencing table
        where rt.fk_Parent = old.pk;
      declare rkey char(2);
      open c;
    l: loop
        fetch next from c into :rkey;
        if (sqlstate <> "00000") then leave l;
        update Unit
          set fk_Parent = null
        where current of c;
        begin
          declare undo handler for binding_exc
            begin end;
            delete from Unit  where current of c;
        end
      end loop;
      close c;
    end
  end;
create trigger T_Delete_Employee
  after delete on Employee
  referencing old row as old for each row
  begin atomic
    declare binding_exc condition;
    if (NOT(old.fk_Carpool_rO is null)) then
      if ((select count(*) from Employee
            where fk_Carpool_rO =
                    old.fk_Carpool_rO) < 2) then
        delete from Carpool where Carpool.pk =
                                old.fk_Carpool_rO;
      end if
    end if;
  end;
create trigger T_Update_fk_Carpool_rO
  after update of fk_Carpool_rO on Employee
  referencing old row as old
  referencing new row as new
  for each row
    when (not(old.fk_Carpool_rO is null) and
        (not(new.fk_Carpool_rO is null) or
        exists(select * from Carpool
                where Carpool.pk =
                  old.fk_Carpool_rO))
  begin atomic
    declare binding_exc condition;
    if ((select count(*) from Employee
          where fk_Carpool_rO =
                  old.fk_Carpool_rO) < 2) then
      delete from Carpool where Carpool.pk =
                              old.fk_Carpool_rO;
    end if
  end;
create trigger T_Update_fk_Parent
  after update of fk_Parent on Unit
  referencing old row as old
  referencing new row as new
  for each row
    when (not(old.fk_Parent is null) and
        (not(new.fk_Parent is null) or
        exists(select * from Unit
                where Unit.pk = old.fk_Parent))
  begin atomic
    declare binding_exc condition;
    if (new.fk_Parent is null) then
      begin
        declare undo handler for binding_exc
          begin end;
        delete from Unit  where Unit.pk =  old.pk;
    end
  end;
```

Fig. 1.23 Partial standard SQL currently needed to implement model shown in Fig. 1.22

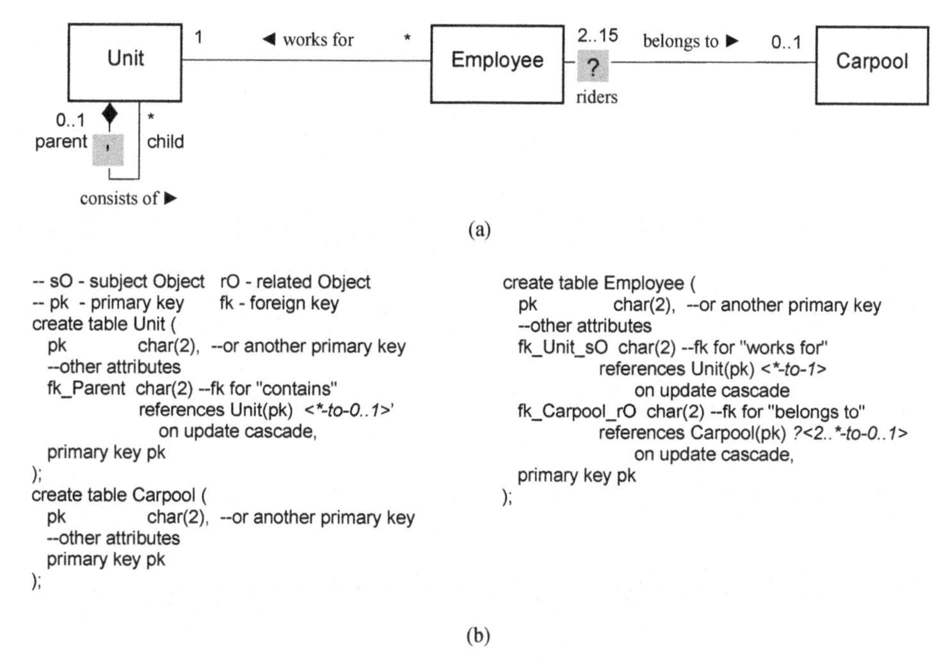


```
-- sO - subject Object   rO - related Object
-- pk  - primary key     fk - foreign key
create table Unit (
  pk          char(2),  --or another primary key
  --other attributes
  fk_Parent  char(2) --fk for "contains"
              references Unit(pk)  <*-to-0..1>'
                on update cascade,
  primary key pk
);
create table Carpool (
  pk          char(2),  --or another primary key
  --other attributes
  primary key pk
);

create table Employee (
  pk            char(2),  --or another primary key
  --other attributes
  fk_Unit_sO  char(2) --fk for "works for"
              references Unit(pk) <*-to-1>
                on update cascade
  fk_Carpool_rO  char(2) --fk for "belongs to"
              references Carpool(pk) ?<2..*-to-0..1>
                on update cascade,
  primary key pk
);
```

(b)

Fig. 1.24 For part of a company database, (a) an extended class diagram and (b) an extended SQL implementation

Chapter 2

Object Relationship Notation (ORN)

The *Object Relationship Notation (ORN)* was first described in Ehlmann et al. (1992). It was proposed for "representing non-inheritance relationships in an object-oriented, scientific database," as indicated by the title of this 1992 paper. Like the relational and ER models, ORN has evolved from what was originally proposed. Its syntax, graphical representation, and semantics have changed slightly, and it has been integrated into UML class diagrams, adapted to relational databases, and applied to the development of all types of databases, not just scientific (Ehlmman and Riccardi 1994, Ehlmann and Stewart 1997, Neal and Ehlmann 2000, Ehlmann and Yu 2002, Ehlmann 2006).

Today, ORN can best be described as a declarative scheme that adds referential actions to UML multiplicities to allow the semantics of associations to be better defined. Its symbols for describing referential actions, called *bindings*, are included in ER and class diagrams to better describe associations during conceptual modeling. Its complete syntax, including multiplicities, is included in DDLs to better define associations to DBMSs, object or relational. This allows a DBMS to enforce multiplicities and perform the specified referential actions. When ORN is supported by an *ORN-extended DBMS*, a variety of association types can be easily modeled, more directly mapped to a database definition, and automatically implemented.

This chapter defines the syntax of ORN, its graphical representation in an ER and class diagram, and its semantics. It explains these semantics with a number of examples and concludes by returning to the company database model given at the end of the previous chapter to explain the ORN previewed by this example. The chapter can be read with Chapter 3, which presents a tool that is helpful in learning ORN.

2.1 Syntax

The syntax of ORN is that of an <*association*> and is given by the syntax diagram in Fig. 2.1. Valid syntax results only from traversing the diagram in the direction of the arrows. Some examples of <*association*>s are given below. As we shall see in Section 2.3, the last two examples are semantically equivalent.

<1-to-*>   '<5..*-to-*>X-   <0..1-to-2..15>~?   <0..1-to-2..15>|?X?

Associations are described in ORN at two levels of detail. A <*multiplicities*> specification describes a binary association type solely by multiplicities, e.g., 1-to-

B.K. Ehlmann, *Object Relationship Notation (ORN) for Database Applications*,
Advances in Database Systems 39, DOI 10.1007/978-0-387-09554-7_2,

1..*, and reflects what is given in a UML class diagram. A *<multiplicity>* is given for each class or role. An *<association>* delimits a *<multiplicities>* specification by < and > symbols and adds more semantic detail by including a *<binding>* for each end of the association, e.g., !<1-to-1..*>?.

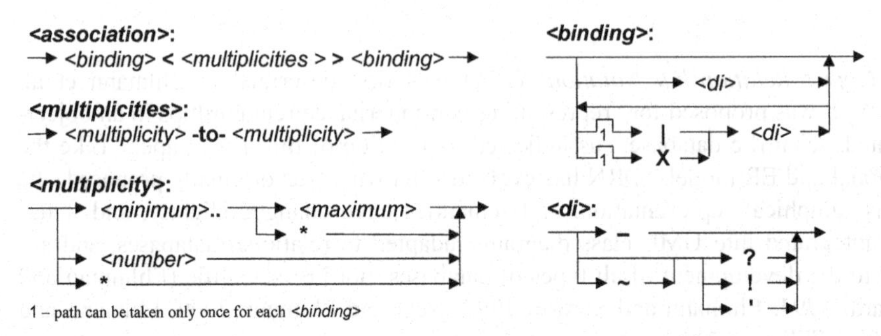

Fig. 2.1 ORN Syntax (Reprinted, with permission, from [Ehlmann 2007] © 2007 IEEE)

A *<binding>* uses just a few special characters, or symbols, to indicate the destructibility of association links, which prescribe one or more referential actions. A *<binding>* is nil, i.e., contains no symbols, which specifies *default binding*; or it is a *<di>*, *destructibility indicator*; or it is a |, an *implicit indicator*, or X, an *explicit indicator*, followed by a destructibility indicator; or it is the latter followed by the opposite of the given implicit or explicit indicator followed by another destructibility indicator. A destructibility indicator is a -, ?, !, or ' symbol, where the latter three symbols can be preceded by an optional ~, a *cascade indicator*.

2.2 Graphical Representation

The graphical representation of ORN makes a modest extension to the ER diagram and class diagram. Essentially a *<binding>*, which may be nil, is included at each end of a binary association. This can be done in a number of ways.

Fig. 2.2 (a) shows how *<binding>*s can be included as stereotype icons in an *ORN-extended class diagram*. In UML, a stereotype icon can be used to extend the standard notation (OMG 2005). The graphical representation of an *<association>* as given in Fig. 2.2 (a) is equivalent to the syntactic representation

<binding1> < *<multiplicity1>* **-to-** *<multiplicity2>* > *<binding2>*

where *<binding1>* and *<multiplicity1>* are given for the *subject class* and *<binding2>* and *<multiplicity2>* are given for the *related class*. An example of the graphical representation for a <0..1-to-2..15>|?X? association is given in Fig. 2.2 (b).

Fig. 2.3 (a) shows the |?X? binding properly placed on the association line in an *ORN-extended ER diagram*, and Fig. 2.3 (b) shows its equivalent placed next to the entity set mapping. The former placement is appropriate when explicit and implicit indicators are given. It is also the original graphical representation for ORN in ER diagrams and is the basis for the ORN logo (see Chapter 3, bottom-left of Fig. 3.1).

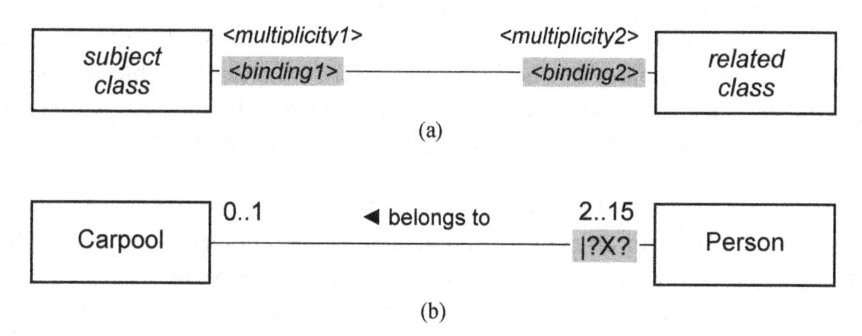

Fig. 2.2 ORN in a class diagram, (a) generic description and (b) example

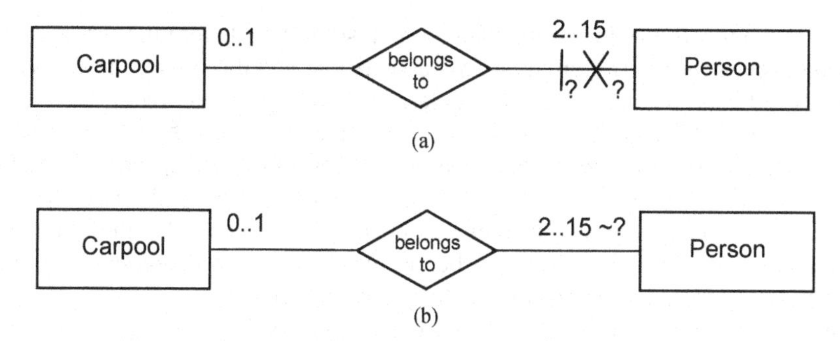

Fig. 2.3 ORN in an ER diagram, symbols placed (a) on the association line and (b) by the mappings

The class diagram representation for an *<association>* as described in Fig. 2.2 is used exclusively in the remainder of this book.

2.3 Semantics

ORN describes the semantics—or one could say the "nature" or "behavior"—of a large variety of binary association types. The semantics of ORN itself are derived from the semantics of the multiplicities and bindings given in an *<association>*. The role of multiplicities in describing association semantics is well known and is described in Chapter 1. The role of "bindings" needs a lot of explaining.

Bindings can be viewed three ways at three different levels.

- At the highest level, bindings when added to multiplicities determine the extent and cohesiveness of complex objects. As we saw in Chapter 1, both the relational and object models disassemble complex objects to some degree. ORN is the "glue" that identifies and holds these objects together. The glue can provide a weak bond, a super bond as for composite objects, or somewhere in between.

- Bindings indicate the degree of "binding" between related objects by indicating the destructibility of association links, both implicit and explicit. Here, *implicit* means automatic, or system initiated, and *explicit* means application initiated, either by a program or query. *Implicit destructibility* describes whether association links involving an object can be implicitly destroyed when an object is deleted and whether such destruction requires related objects to be deleted. *Explicit destructibility* describes similar rules that apply when association links are explicitly destroyed.
- At the lowest level, bindings indicate the referential actions, or *multiplicity actions*, that should occur on object deletion and association link destruction. These system actions enforce multiplicity integrity (not just referential integrity), provide the desired implicit and explicit destructibility for an association, and implement the required implicit cascading of link destruction and deletion operations within a complex object operation.

In an *<association>*, the *<multiplicity>* and *<binding>* before the -to- apply to the subject class, or role for an intra-class association, and those after the -to- apply to the related class or role. The subject class (or role) can be viewed as the related class (or role) and vice versa, in which case the inverse *<association>* applies. In general, the inverse of $b_1<m_1\text{-to-}m_2>b_2$ is $b_2<m_2\text{-to-}m_1>b_1$, and specifically that of <0..1-to-2..15>~? is ~?<2..15-to-0..1>. The *<multiplicity>* and *<binding>* semantics for one end of an association are independent of those for the other end. Those for the other end, however, can be "in play" on certain operations, e.g., link destruction, and so can affect the final outcome of an operation.

Table 2.1 gives the semantics of ORN from the perspective of a subject class S in an association A. We shall refer to this table often as we study examples of ORN that illustrate the different bindings. One thing to keep in mind with ORN semantics is that object deletion and a link destruction, implicit or explicit, can actually be implemented as object archival then deletion and link archival then destruction. In many databases today, it is rare to actually delete or destroy anything.

2.4 Examples

To illustrate how ORN describes association semantics, this section presents variations on the classic, normally one-to-many association between departments and employees. Instead of using "department," however, I use "unit" since it is more generic (and shorter in length). So, the association we now focus on is: an employee "works for" a unit. Although the semantics of some of the variations on this association that we examine will seem odd, they make sense for other associations, which we explore later in this chapter and Chapter 4. Here, I focus on just this one association so the reader can focus solely on the varied multiplicities and bindings and not have to think about new associations and their own semantic peculiarities.

Table 2.1 ORN Semantics

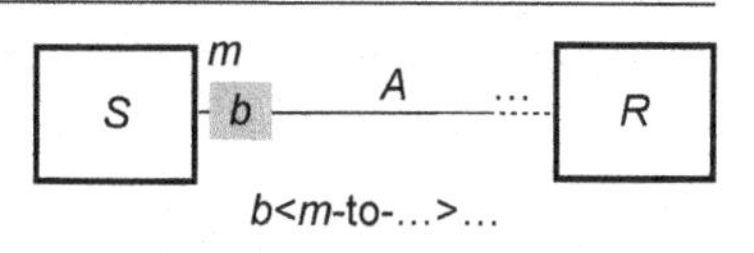

Semantics are given in terms of a subject class *S* with multiplicity *m* and binding *b* in an association *A* with some related class *R* (which could be *S* in a different role).

***<multiplicity>*:** Semantics are the same as those in UML. Essentially, *m* indicates a lower bound and upper bound on the number of objects of type *S* that can be related via *A* to each object of type *R*.

An *R* object can be created provided this does not violate *m*. The check for a lower bound violation is deferred until transaction commit. An *A* link can be created provided this does not violate *m*. The check for an upper bound violation is immediate. The enforcement of *m* on the deletion of an *S* object or destruction of an *A* link is determined by the binding *b*.

***<binding>*:** A **|** in *b* denotes a "cut" and an **I**mplicit, i.e., system initiated, destruction of an existing *A* link that must occur on deletion of an *S* object. An **X** in *b* denotes a "cross out" and an e**X**plicit, i.e., user initiated, destruction of an *A* link.[1]

An *S* object deletion and an explicit *A* link destruction are *complex object operations*. Deletion of an *S* object succeeds only if all existing association links involving that object are implicitly destructible. Also, deletion of an *S* object or explicit destruction of an *A* link succeeds only if all required implicit object deletions succeed.

***<di>*:** A **d**estructibility **i**ndicator in *b* specifies the destructibility of an *A* link. The meaning of each indicator is given below. This meaning can alternatively be described by the actions taken on an attempt to destroy an *A* link. These actions are given in brackets. If a *<di>* is given after a **|**, it applies to implicit link destruction; if given after an **X**, it applies to explicit link destruction; and if given alone, it applies to both. If a *<di>* is not given, i.e., is nil, for implicit link destruction, explicit link destruction, or both, default destructibility applies to whichever.

nil — *Default destructibility.* A link can be destroyed provided this does not violate *m*.[2] [Destroy the link. If *m* is violated[2], raise an exception[3].]

- — *Negative destructibility.* A link cannot be destroyed. [Raise an exception.[3]]

~? or ? — *Conditional cascade destructibility.* A link can be destroyed, but if this violates *m* (**?**), the destruction must be cascaded (**~**) to the related *R* object, i.e., this object must be implicitly deleted. [Destroy the link. Delete the related *R* object**?** If *m* is violated, yes; else no.]

~! or ! — *Emphatic cascade destructibility.* A link can be destroyed, but the destruction must be cascaded (**~**) to the related *R* object. [Destroy the link. Delete the related *R* object**!**]

~' or ' — *Tentative (or qualified) cascade destructibility.* A link can be destroyed, but an attempt must be made to cascade (**~**) the destruction to the related *R* object; however, this implicit *R* object deletion must be undone if it fails, but is required if and only if its undoing would violate *m*.[2] (Think of the **'** as a "pruned back **!**" or as a "qualifying footnote reference" on the cascade.) [Destroy the link. Delete the related *R* object**'** (**'** – If an exception occurs on this nested complex object operation, undo the delete and then, if *m* is violated[2], raise an exception[3].)]

1 - A link change done as a single operation that replaces the *S* object with another is not treated as an explicit link destruction relative to class *S* (but is relative to *R*) and is allowed if allowed by other multiplicities and bindings.
2 - The check for a lower bound violation is deferred until the end of the current complex object operation.
3 - The current complex object operation is undone.

The generic class diagram for the "works for" association is shown in Fig. 2.4 with variables given for the multiplicities and bindings. Also, included in the figure is the syntactical representation of the *<association>*, given in terms of these variables. The examples discussed in this section are created by assigning different multiplicities and bindings to these variables.

In discussing the variations on the "works for" association, we derive the precise meaning of each binding by paraphrasing from Table 2.1 while making appropriate variable substitutions. A is always "works for." S = Unit, R = Employee, $m = m_U$, and $b = b_U$ to derive the meaning of a Unit binding. To derive the meaning of an Employee binding, S = Employee, R = Unit, $m = m_E$, and $b = b_E$.

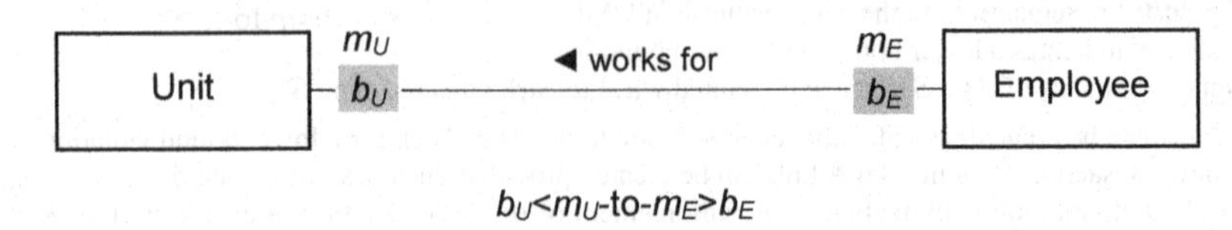

Fig. 2.4 Class diagram for "works for" association with variable multiplicities and bindings

2.4.1 <*-to-*>

This *<association>* imposes no constraints on the "works for" association. A unit can contain none or infinitely many employees, and each employee can belong to none or infinitely many units. The bindings for both classes are default since no binding symbols are given The class diagram is shown in Fig. 2.5 (which is the diagram in Fig. 2.4 where $m_U = m_E$ = * and $b_U = b_E$ = nil).

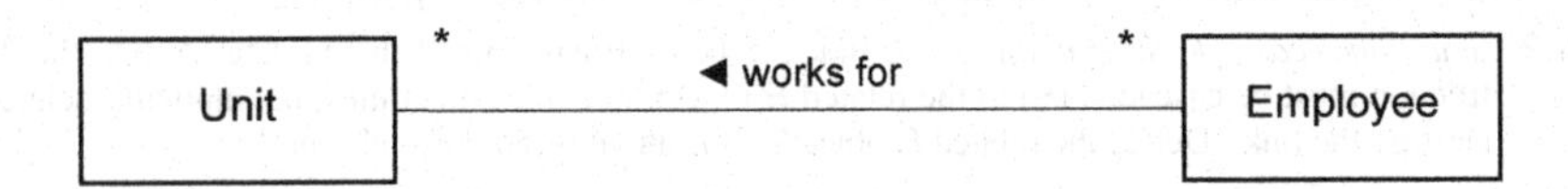

Fig. 2.5 Class diagram for a <*-to-*> "works for" association

The *default binding* for Unit implies both *implicit default binding* and *explicit default binding*. ORN semantics require that an object cannot be deleted unless all associations that it has with other objects can be implicitly destroyed. This means that when an object of type Unit is deleted, all links that it has with other objects, including any Employee objects, must be implicitly destructible. The implicit binding for a deleted object's class, e.g., Unit, in an association, e.g., "works for," determines whether such implicit destruction of the links of the association type is allowed.

Thus, the implicit default binding for the Unit class in the "works for" association applies when a Unit object is deleted and means (paraphrasing from the nil case for a destructibility indicator in Table 2.1): a "works for" link can be destroyed provided this does not violate the * multiplicity. A *, i.e., 0..*, multiplicity is never violated. Thus, a unit can always be deleted, even if it has some employees (unless the deletion is disallowed by other associations in which it is involved). The referential actions given for the default binding in Table 2.1 make it clear (if it was not already) that deletion of a unit results only in the implicit destruction of the "works for" links with related employees; the related employees are retained.

The explicit default binding for Unit applies when a "works for" link is explicitly destroyed and means (again, paraphrasing from Table 2.1): the link can be destroyed provided this does not violate the * multiplicity. Again, a * is never violated. Thus, a "works for" link can always be explicitly destroyed, at least based on the Unit multiplicity and binding. For explicit destruction, however, the multiplicities and bindings at both ends of an association are applicable.

The implicit default binding for the Employee class in the "works for" association applies when an Employee object is deleted and means (paraphrasing from Table 2.1, now where S = Employee and R = Unit): a "works for" link can be destroyed provided this does not violate the * multiplicity. A * is never violated. Thus, an employee can always be deleted.

The explicit default binding for Employee applies when a "works for" link is explicitly destroyed and means (paraphrasing from Table 2.1): the link can be destroyed provided this does not violate the * multiplicity. A * is never violated. Thus, a "works for" link can always be explicitly destroyed, based on the multiplicities and explicit bindings at both ends of the association.

2.4.2 <1-to-*>

This *<association>* may be the most appropriate description for the "works for" association. It is the one shown in the class diagram of Fig. 1.22. A unit can contain any number of employees but each employee must belong to one and only one unit. The association exemplifies the use of a default binding with a constrained multiplicity lower bound and upper bound. The class diagram is shown in Fig. 2.6 (which is the diagram in Fig. 2.4 where $m_U = 1$, $m_E = *$, and $b_U = b_E =$ nil).

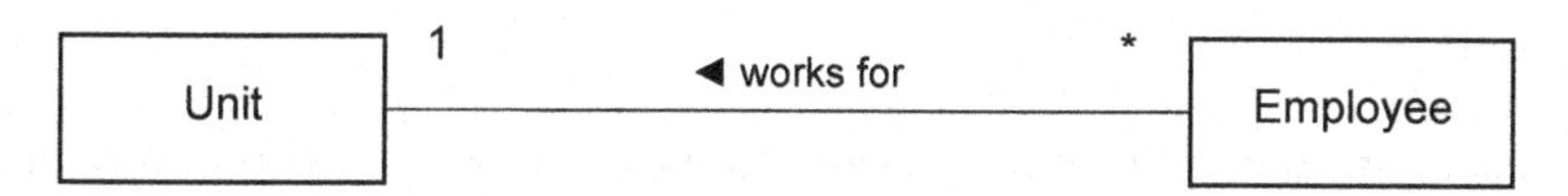

Fig. 2.6 Class diagram for a <1-to-*> "works for" association

The 1, or more precisely the 1..1, multiplicity for Unit in this association imposes constraints on object and link creation. Paraphrasing the second paragraph under ***<multiplicity>*** in Table 2.1: an Employee object can be created provided this does not violate the 1 multiplicity. The check for a lower bound violation (of 1) is *deferred* until transaction commit (i.e., delayed until the commit of the application-defined transaction). A "works for" link can be created provided this does not violate the 1 multiplicity. The check for an upper bound violation (of 1) is *immediate* (i.e., done as part of the current operation). Thus, an employee cannot be created unless it is linked to a unit, and an employee cannot be linked to more than one unit.

The implicit default binding for the Unit class in the "works for" association applies when a Unit object is deleted and means (paraphrasing from Table 2.1): a "works for" link can be destroyed provided this does not violate the 1 multiplicity. The 1 is violated if the Unit object being deleted is linked to an Employee object because an employee must belong to a unit. Thus, a unit cannot be deleted if it has any employees. It can, however, be deleted if it has no employees (at least based on this association).

The explicit default binding for Unit applies when a "works for" link is explicitly destroyed and means (paraphrasing from Table 2.1): the link can be destroyed provided this does not violate the 1 multiplicity. Here again, the 1 is violated because an employee must belong to a unit. Thus an association between a unit and an employee, once created, cannot be explicitly destroyed.

It can, however, be changed. Paraphrasing the fine print of footnote 1 in Table 2.1: A link change done as a single operation that replaces the Unit object with another is not treated as an explicit link destruction relative to the Unit class (but is relative to the Employee class) and is allowed if allowed by other multiplicities and bindings. We shall soon see that this link change is allowed by the multiplicity and binding of the Employee class, for which the link change **is** treated as an explicit link destruction (since a unit is losing an employee and another is gaining one). Thus an employee who works for a unit can be assigned to another unit.

The implicit and explicit default bindings for the Employee class mean the same as they did in the <*-to-*> association. Explicit destruction of a "works for" link is allowed, at least based on * multiplicity and default binding for Employee. While this allows the link change discussed in the previous paragraph, explicit destruction of a "works for" link is stymied by the 1 multiplicity and default binding at the Unit end of the association. By the way, a link change that exchanges one employee for another is disallowed (see again footnote 1). Allowing such would mean an employee would be left without a unit.

Since both footnotes 2 and 3 of Table 2.1 are referenced for default destructibility and are relevant to a 1 multiplicity, I now explain the last of the fine print. I do so in the context of the <1-to-*> "works for" association and an attempted deletion of the Unit object.

With ORN an explicit object deletion is a *complex object operation*, as is any explicit link destruction or change. Such an operation takes place within its own internal transaction and can involve many implicit link destructions and object deletions based on the ORN semantics of one or more associations. For a Unit object deletion, footnote 2 means that the check for a lower bound violation of the 1, i.e., 1..1, multiplicity is not done until the end of the transaction that encompasses this complex object operation, i.e., it is partially deferred. In most cases, deferred versus immediate checking is immaterial in understanding ORN semantics, but in some unusual cases, it can make a difference. For example, in deleting a Unit object it is possible—though very unlikely with this association—that because of other associations and bindings, an Employee object that was linked to the Unit object at the beginning of the complex object deletion has been implicitly deleted by the time of its conclusion. If so, deferred checking at the commit of the operation will not reveal a violation of the 1 multiplicity for this deleted Employee object. Thus, its original link to the deleted Unit object will not cause the failure of the explicit Unit object deletion.

Footnote 3 states that the complex object operation is undone, which means the internal transaction that encompasses the operation is rolled back. Thus, when the deletion of a Unit fails because, for instance, the 1 multiplicity for Unit in the "works for" association is violated at the end of this complex operation, all implicit link destructions and object deletions resulting from the operation, as well as the original explicit Unit deletion itself, are rolled back.

As we have seen, the semantics of default bindings simply enforce multiplicities. Next, we examine the non-default bindings.

2.4.3 <0..1-to-*>|-

With this *<association>* a unit can have none or many employees and each employee may (or may not) work for a single unit. An implicit negative destructibility binding is given for Employee. The class diagram is shown in Fig. 2.7 (which is the diagram in Fig. 2.4 where m_U = 0..1, m_E = *, and b_U = nil, b_E = |-).

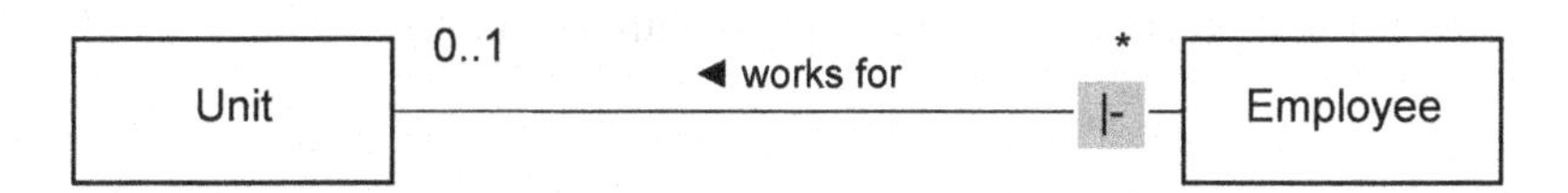

Fig. 2.7 Class diagram for a <0..1-to-*>|- "works for" association

The |- binding for Employee in the "works for" association applies when an Employee object is deleted and means (paraphrasing from Table 2.1): a "works for" link cannot be destroyed. Thus, since ORN dictates that an object cannot be deleted unless all existing links involving the object can be implicitly destroyed, an employee who works for a unit cannot be deleted. Such an employee's link with a unit would have to be explicitly destroyed, which is now allowed by the 0..1 multiplicity and explicit default binding, before the employee could be deleted.

2.4.4 <0..1-to-*>|-X-

This *<association>* is like the previous one except that an explicit negative destructibility binding is given for the Employee class. Since both implicit and explicit negative destructibility are specified, the association can also be described as <0..1-to-*>-. Its class diagram is shown in Fig. 2.8 (which is the diagram in Fig. 2.4 where m_U = 0..1, m_E = *, and b_U = nil, b_E = |-X- or -).

The X- binding for Employee in the "works for" association applies when a "works for" link is explicitly destroyed and means (from Table 2.1): a "works for" link cannot be destroyed. Thus, an association between a unit and an employee, once created, cannot be explicitly destroyed.

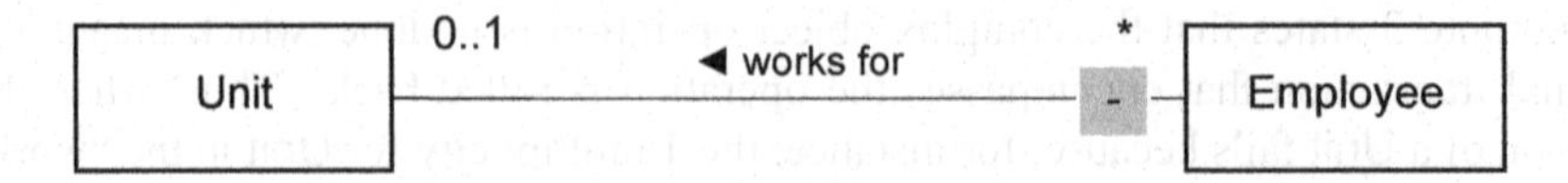

Fig. 2.8 Class diagram for a <0..1-to-*>- "works for" association

It can, however, be changed. For the X- binding, the variable *S* is Employee and *R* is Unit, and we again paraphrase footnote 1 in Table 2.1: a link change done as a single operation that replaces the Employee object with another is not treated as an explicit link destruction relative to the Employee class (but is relative to the Unit class) and is allowed if allowed by other multiplicities and bindings. This link change is allowed by the multiplicity and binding of the Unit class, for which the link change is treated as an explicit link destruction. Thus an employee who works for a unit can be replaced by another employee.

This link change would not be allowed if we add the X- binding to the Unit end of this association. An X-<0..1-to-*>- association joins an employee with a unit until "death do ye part." Here, this means until the unit is deleted because the |- binding for Employee keeps a linked employee from being deleted. (The link change not allowed by the X-<0..1-to-*>- association would also not be allowed by a <1-to-*>- association, because an employee would be left without a unit.)

2.4.5 <0..1-to-1..>?*

This <*association*> specifies that each unit must have at least one worker. It exemplifies the use of conditional cascade destructibility with a lower bound multiplicity constraint. Since both |? and X? are implied by the ?, the association can also be described as <0..1-to-1..*>|?X?, or <0..1-to-1..*>|~?X~? if the optional cascade symbol is given. The class diagram for this association is shown in Fig. 2.9.

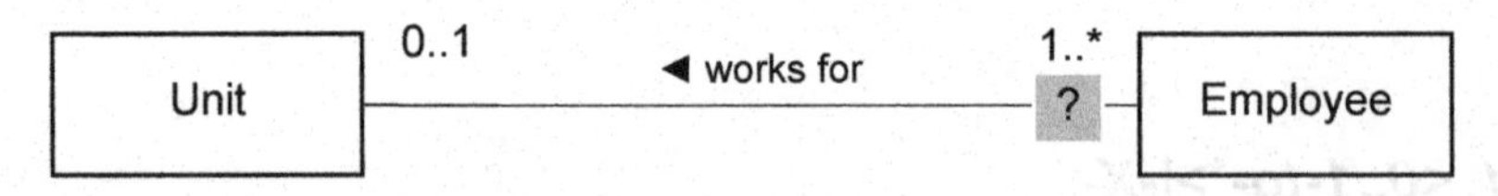

Fig. 2.9 Class diagram for a <0..1-to-1..*>? "works for" association

The |? binding for the Employee class in the "works for" association applies when an Employee object is deleted and means (paraphrasing from Table 2.1): a "works for" link can be destroyed, but if this violates the multiplicity 1..*, the destruction must be cascaded to the related Unit object, i.e., this object must be implicitly deleted. The 1 is violated if the Employee object being deleted is the only one linked to the related Unit object. Thus, deletion of the last employee who works for a unit causes the deletion of the unit.

The X? binding applies when a "works for" link is explicitly destroyed and means (from Table 2.1): a "works for" link can be destroyed, but if this violates the multiplicity 1..*, the destruction must be cascaded to the related Unit object, i.e., this object must be implicitly deleted. Again, the 1 is violated when the Employee object being "de-linked" from the Unit object is the only one linked to this object. Thus terminating the "works for" relationship between a unit and its last employee terminates the unit!

If the association was <0..1-to-2..*>?, then deletion or explicit delinking of one of just two employees working for a unit would cause deletion of the unit. Assuming this association, Fig. 2.10 shows object diagrams representing the state of a database before and after the deletion of an employee e4. In an object diagram, circles represent objects and lines represent the association links between these objects.

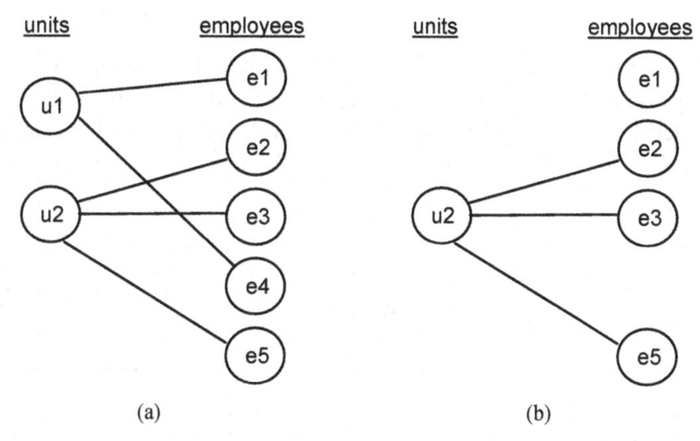

Fig. 2.10 For a <0..1-to-2..*>? association between units and employees, object diagrams for the database state (a) before and (b) after deletion of e4

2.4.6 !<0..1-to->*

This *<association>* exemplifies the use of emphatic cascade destructibility. Since both |! and X! are implied by the !, the association can also be described as |!X!<0..1-to-*>, or |~!X~!<0..1-to-*> if the optional cascade symbol is given. The class diagram for this association is shown in Fig. 2.11.

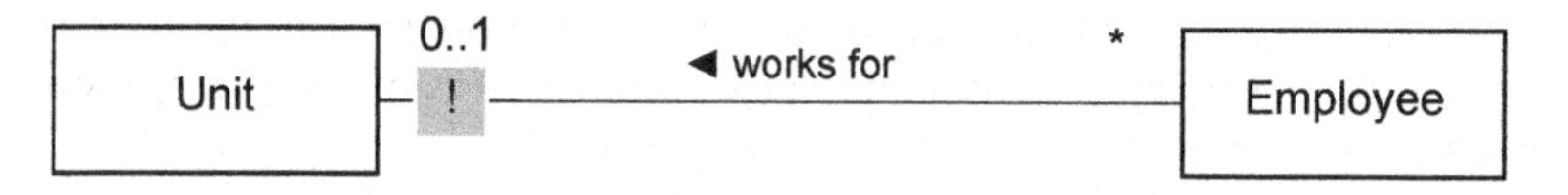

Fig. 2.11 Class diagram for a !<0..1-to-*> "works for" association

The |! binding for the Unit class in the “works for” association applies when a Unit object is deleted and means (from Table 2.1): a “works for” link can be destroyed, but the destruction must be cascaded to the related Employee object (i.e., this related object must be implicitly deleted). Thus, deletion of a unit causes the implicit deletion of all employees who work for that unit.

The X! binding applies when a “works for” link is explicitly destroyed and means (from Table 2.1): the “works for” link can be destroyed, but the destruction must be cascaded to the related Employee object (i.e., this related object must be implicitly deleted). Thus terminating the “works for” relationship between a unit and an employee always terminates the employee.

2.4.7 '<-to-1..*>*

This <*association*> exemplifies the use of tentative cascade destructibility. It allows an employee to work for many units but each unit must have at least one employee. Since both |' and X' are implied by the ', the association can also be described as |'X'<*-to-1..*>. The class diagram for this association is shown in Fig. 2.12.

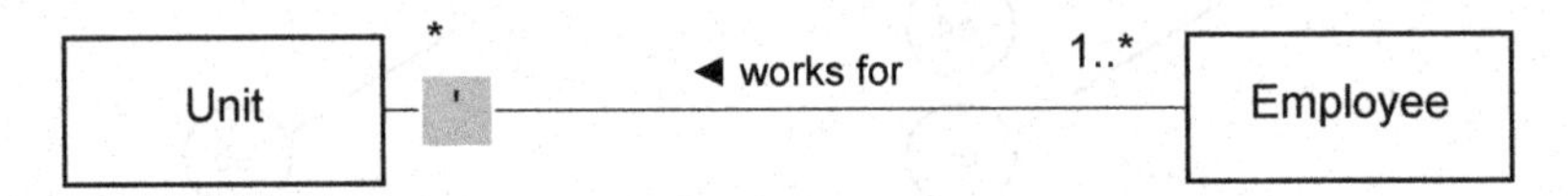

Fig. 2.12 Class diagram for a '<*-to-1..*> “works for” association

The |' binding for the Unit class in the “works for” association applies when a Unit object is deleted and means (from Table 2.1): a “works for” link can be destroyed, but an attempt must be made to cascade the destruction to the related Employee object; however, this implicit Employee object deletion must be undone if it fails, but is required if and only if its undoing would violate the * multiplicity. A * is never violated, so the deletion of the related Employee object is never required. Here, the attempted deletion of a related object fails if it is the only Employee object related to another Unit object. It fails because of the lower bound 1 multiplicity and default implicit binding for the Employee class, which does not permit the implicit destruction of an **existing** link when the multipicity is violated. Thus, deletion of a unit causes the implicit deletion of all employees who work for that unit except any such employee who is the only employee working for another unit.

The X' binding applies to the explicit destruction of a “works for” link. Its meaning is similar to that of the |' binding. Essentially, on explicit destruction of a link, the related Employee object is implicitly deleted unless it is the only Employee object linked to another Unit object.

Let us suppose that on deletion of a unit or explicit destruction of an employee’s working relationship with a unit, we do not want to delete **any** employee who is working for another unit. Then, the <association> should be '<*-to-1..*>|-. Now, the |- binding for the Employee class blocks the implicit destruction of an existing link on an attempted deletion of an employee, thus making it fail. Employees, however, who do not have existing links with other units are deleted. Assuming this association, Fig. 2.13 shows object diagrams representing the state of a database before and after the deletion of a unit u3. The "blockage" provided by the |- is independent of the Employee multiplicity and so would also work for a '<*-to-*>|- association between units and employees.

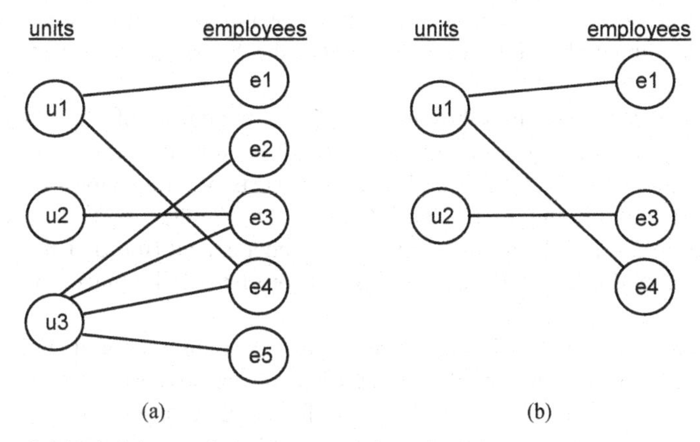

Fig. 2.13 For a '<*-to-1..*>|- association between units and employees, object diagrams for the database state (a) before and (b) after deletion of u3

2.5 Flashback to the Company Database

We now revisit the company database example given at the end of Chapter 1. Fig. 2.14 shows the same extended class diagram that was given in Fig. 1.24 (a). The extensions to the diagram—i.e., the ORN bindings, both explicit and default—can now be examined and understood based on ORN semantics presented in this chapter.

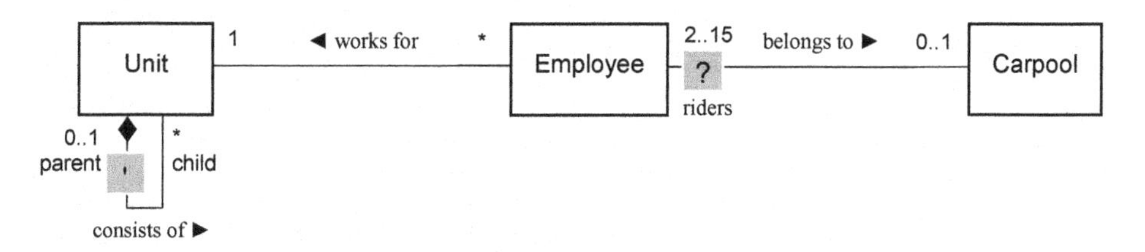

Fig. 2.14 ORN-extended class diagram for part of a company database

Repeated below are the descriptions of relevant association semantics for this application that could not be modeled in the standard class diagram. The applicable bindings and multiplicities that represent each semantic in the extended-ORN class diagram are given within parentheses. Also given within these parentheses is a reference to a subsection of Section 2.4 that describes a semantically similar variation of the "works for" association between employees and units. Hopefully now, after a bit of study, how these nontrivial semantics can be represented in a class diagram by simply adding two symbols—' and ?—will be clear to the reader.

- If an attempt is made to explicitly remove a unit that has employees working for it, an error should result because all employees must work for a unit (the implicit default binding and 1 multiplicity for Unit in the "works for" association, see Section 2.4.2).
- If a unit is removed, all subordinate, i.e., component or child, units should be automatically removed unless those units still have employees working for them (the |' binding for the parent end of the "consists of" association along with the implicit default binding and 1 multiplicity for Unit in the "works for" association, see Section 2.4.7). Such units may remain independent (the 0..1 multiplicity for the parent end of the "consists of" association) or later be placed within another unit.
- If a unit is removed from its parent unit, similar semantics should apply. It and all of its subordinate units should be automatically removed unless they have employees working for them (the X' binding for the parent end of the "consists of" association along with the implicit default binding and 1 multiplicity for Unit in the "works for" association, see Section 2.4.7).
- A carpool is defined by having at least two riders, so that if the number of riders falls below two, either because of employee terminations or because employees quit the carpool, the carpool really no longer exists and so should be automatically removed (the ? binding and 2..15 multiplicity for Employee in the "belongs to" association, see Section 2.4.5).

The next chapter discusses a tool that allows the reader to very easily model all of the associations presented in this chapter using ORN and to readily observe their semantics being automatically enforced by a database system. The visualization provided by this type of tool is intended to help the user become competent at using ORN.

Chapter 3

ORN Simulator

A Modeling Tool Where Associations Come Alive

The ORN Simulator is a Web-accessible, prototype, data modeling tool that allows a user to model a database and then immediately simulate its operation (Ehlmann and Riccardi 1999, Ehlmann 2002). The user defines object types, or classes, and relationships, or associations, in the context of an ORN-extended ER or class diagram. The model defined by the diagram "comes alive" in the sense that the user can readily create, view, and manipulate a prototype database in the context of the model. This allows a database designer to observe, fine-tune, and verify the semantics, or behavior, of associations before implementing them in a real database application.

It also allows a user of the tool to develop a better understanding of ORN and other database management concepts. I developed the ORN Simulator as a research prototype to showcase ORN and the concept of simulating a database within its model. I soon discovered, however, that this tool could be an effective pedagogical tool for learning the concepts of data modeling, transaction processing, and ORN. As such, I have been using the tool, I believe quite successfully, in my software engineering and database management classes since 2001.

The ORN Simulator is a prototype. The number of object and relationship types that can be modeled in one diagram is very limited. To model a typical database, a number of overlapping diagrams would be required. Also, only associations can be modeled, not "is a" relationships. A real modeling tool, perhaps developed based on the concepts inherent in the ORN Simulator, would not have such limitations.

This chapter provides an overview of the ORN Simulator, showing how it can be used to first create a database model and database and then verify association behavior. To do this, it utilizes many screen shots and the company database example used in the first two chapters. The chapter also discusses the architecture of the ORN Simulator and concludes by summarizing the benefits that such a tool can offer.

The ORN Simulator can be accessed via the Web by clicking on the "ORN Simulator" link at www.siue.edu/~behlman. The reader is encouraged to create the sample model and database shown in this chapter and perform the operations demonstrated herein. The reader may also wish to use the ORN Simulator to experiment with and verify the semantics of any of the many associations presented in this book.

3.1 Creating a Database Model and a Database

After clicking on the "ORN Simulator" link, an introductory screen is shown for ORN and the ORN Simulator. After clicking on the "Run ORN Simulator" link on

B.K. Ehlmann, *Object Relationship Notation (ORN) for Database Applications*,
Advances in Database Systems 39, DOI 10.1007/978-0-387-09554-7_3,

this screen, a window appears asking for the user's name and affiliation, which is only used to track who is using the tool. After clicking OK on this screen, the screen shown in Fig. 3.1 appears. It is an ER diagram, but we wish to use a class diagram.

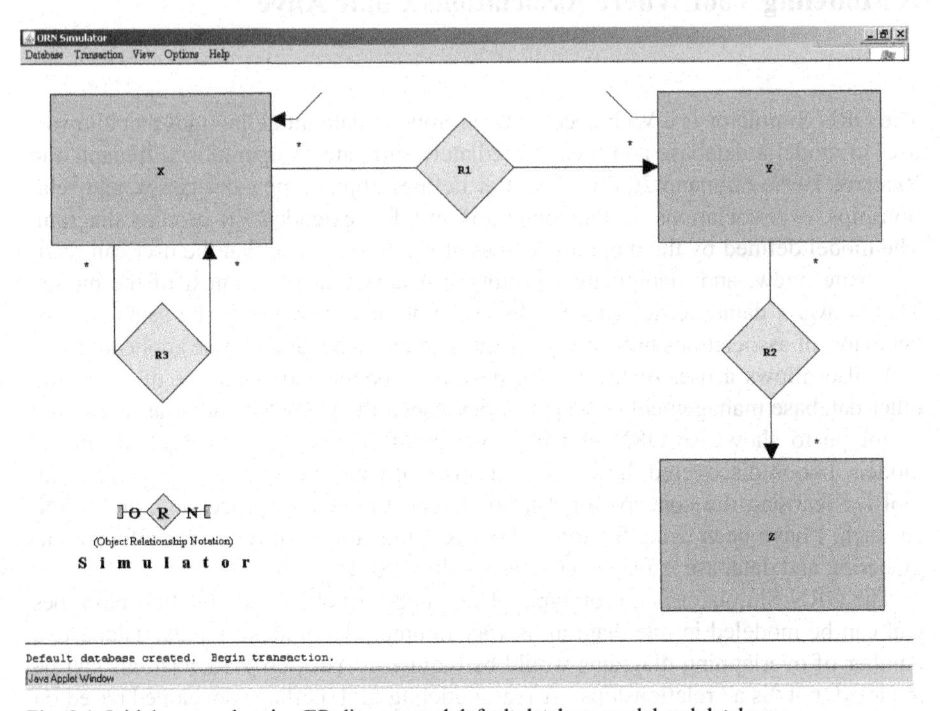

Fig. 3.1 Initial screen showing ER diagram and default database model and database

After clicking on "View" and "UML Class Diagram" as shown in Fig. 3.2, the screen shown in Fig. 3.3 appears, which shows the default model and database. The model contains three classes—X, Y, and Z—and four associations—R1, R2, R3, and R4. The association line for R4 between classes Y and X is only partially shown in Fig. 3.3. Clicking "View" and "R1, R3, R4" results in the screen shown in Fig. 3.4, which fully displays the R4 association. All associations in the default model and database are <*-to-*>. The default database is empty, i.e., there are no objects shown within any class. This will become clear after some objects have been created.

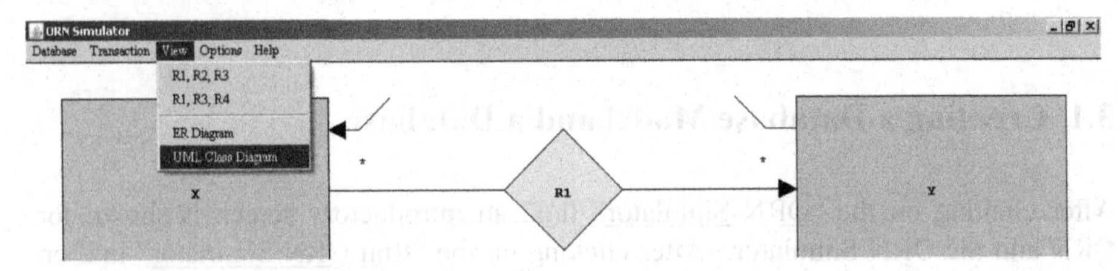

Fig. 3.2 Top of screen showing selection of the UML Class Diagram view

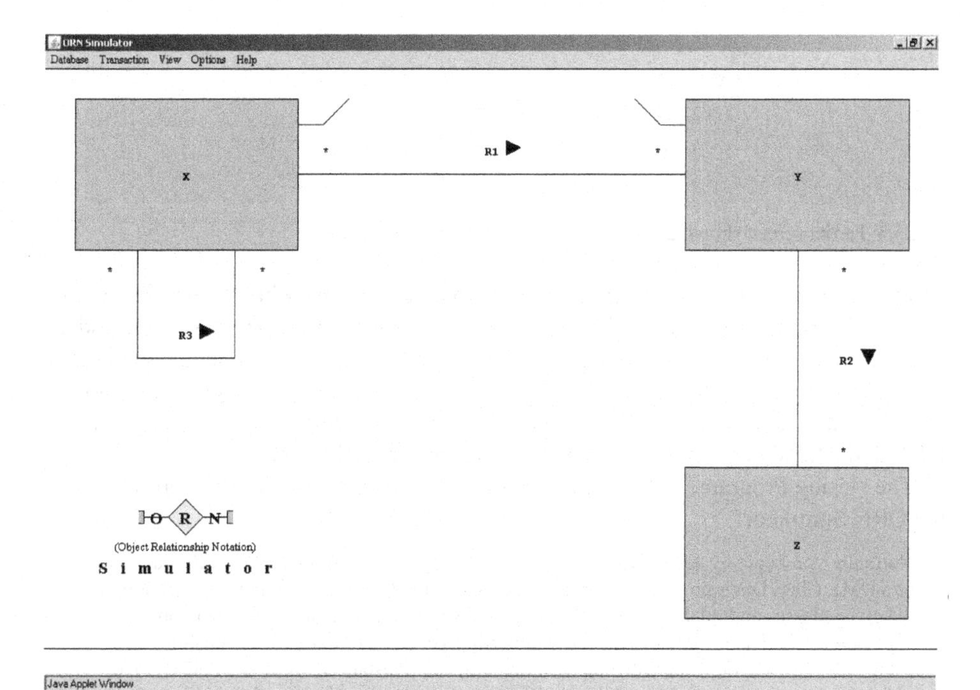

Fig. 3.3 Screen showing UML class diagram and default database model and database

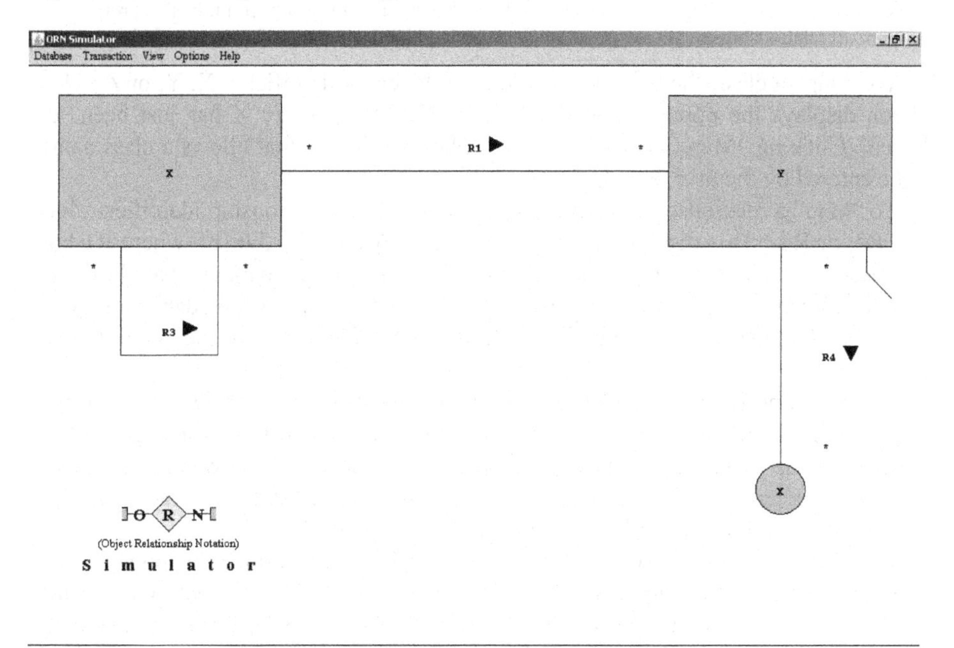

Fig. 3.4 Screen showing association R4

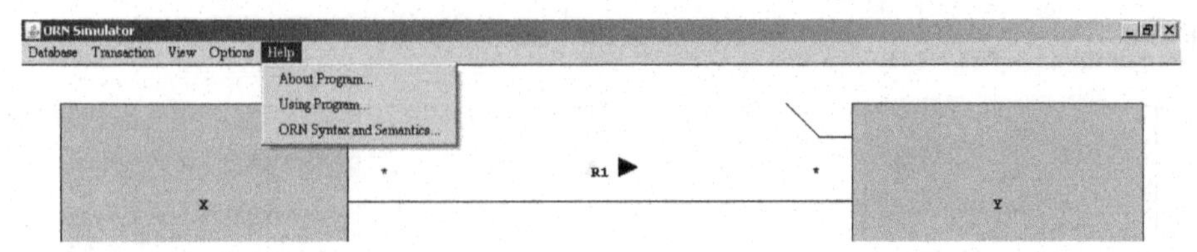

Fig. 3.5 Partial screen showing dropdown menu selections for Help

Fig. 3.5 shows the dropdown menu selections for the menu bar "Help" selection. Clicking "About Program…" displays a window giving the current version and a one-sentence description of the tool. "Using Program…" displays a window that explains how to use the tool, including the functions provided by all menu selections. "ORN Syntax and Semantics…" displays a window showing the syntax diagram for ORN and a table very similar to Table 2.1, describing ORN semantics.

The "Using Program…" help text explains to the user the beginning steps of using the ORN Simulator:

> Normally you begin by selecting the "View" in which you wish to operate, an ER Diagram or a UML Class Diagram. … [We have already done this.] Then you "Map" one or more of the given classes and relationships (UML associations) in the diagram to those in your application … . Next, for each mapped relationship you can "Redefine Relationship Type" using ORN. Now you can create objects and relationship instances (or links), delete objects, and destroy and change links, and thereby observe the behavior of your relationships. If the behavior is not what is desired, you can clear the database, "Redefine Relationship Type," and perform database updates to observe the new behavior. You would continue this process until all relationships in your application have been defined and their behavior verified.

To "Map" a class, the user clicks on one of the class identifiers: X, Y, or Z. This action displays the popup menu shown in Fig. 3.6 (a), where X has just been selected. Clicking "Map Object Class…" displays a window that allows a class name to be entered by the user, e.g., Unit.

To "Map" a relationship, the user clicks on one of the relationship identifiers: R1, R2, R3, or R4. This displays the popup menu shown in Fig. 3.6 (b), where R1 has just been selected. Clicking "Map Relationship…" displays a window that allows an association name to be entered by the user, e.g., "works for." (A readability indicator can be included in the name if readability is opposite to that of the given default direction.)

To "Redefine Relationship Type" for a relationship, the user clicks on one of the relationship identifiers. Again, this displays the popup menu shown in Fig. 3.6 (b). Clicking "Map Relationship Type…" displays a window that allows an <*association*> to be entered by the user, e.g., <1-to-*>. This <*association*> replaces the current one defined for the relationship, e.g., the default <*-to-*>.

In addition to allowing relationships to be mapped and redefined, the popup menu displayed for a relationship also allows role names to be given for both ends of the association. The given readability indicator for a relationship identifier always indicates a reading from subject to related class—e.g., for R1, X is the subject class and Y is the related class.

Once the database model has been defined, objects are created and deleted using the popup menu selections shown in Fig. 3.6 (a), and association links are created, destroyed, or changed using the popup menu selections shown in Fig. 3.6 (b).

Fig. 3.7 shows the company database model that matches Figs. 1.24 (a) and 2.14 after classes and relationships have been properly mapped, relationships have been properly redefined, and role names have been defined. In the context of the model, it also shows a conforming company database after some objects and links have been created. A link, which is an ordered pair, appears as a subject object followed by <--> followed by a related object. That is, `(u0, e0)` is represented as `u0<-->e0`.

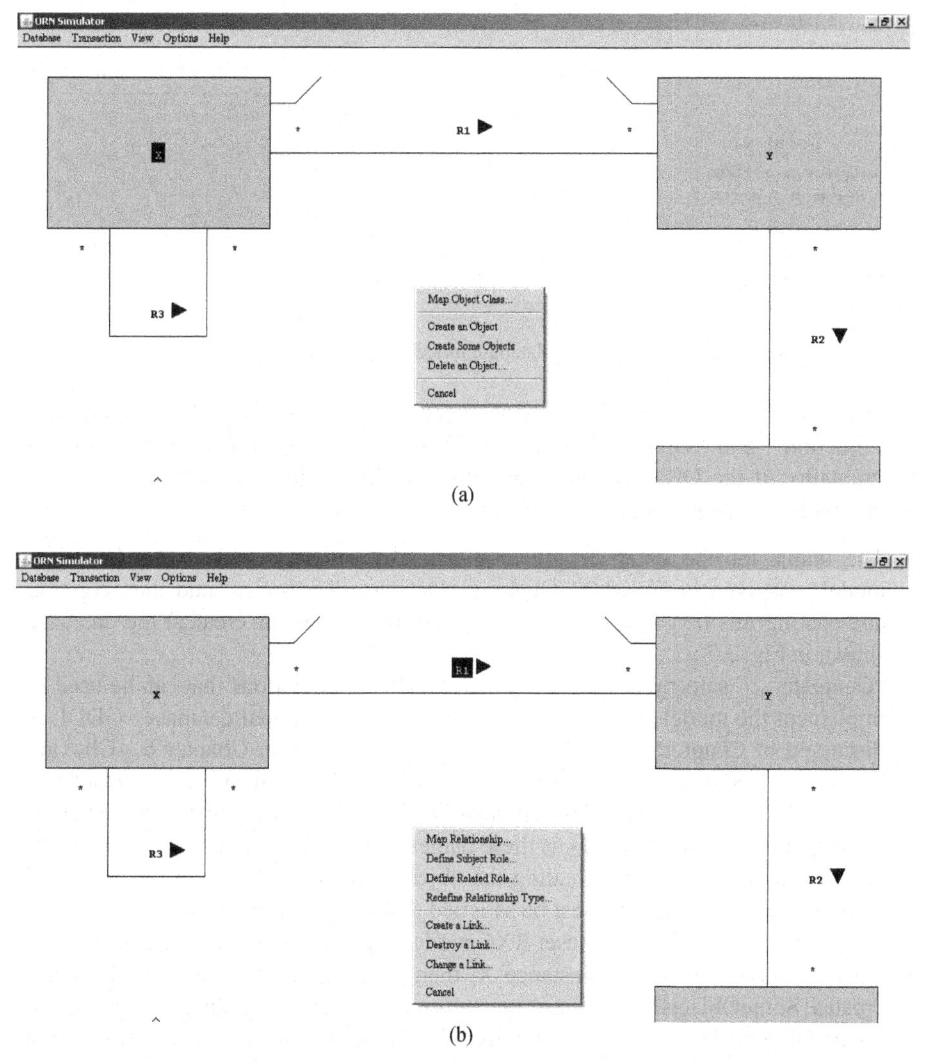

Fig. 3.6 Partial screens showing popup menus that appear (a) after clicking on a class id, e.g., X, and (b) after clicking on a relationship id, e.g., R1.

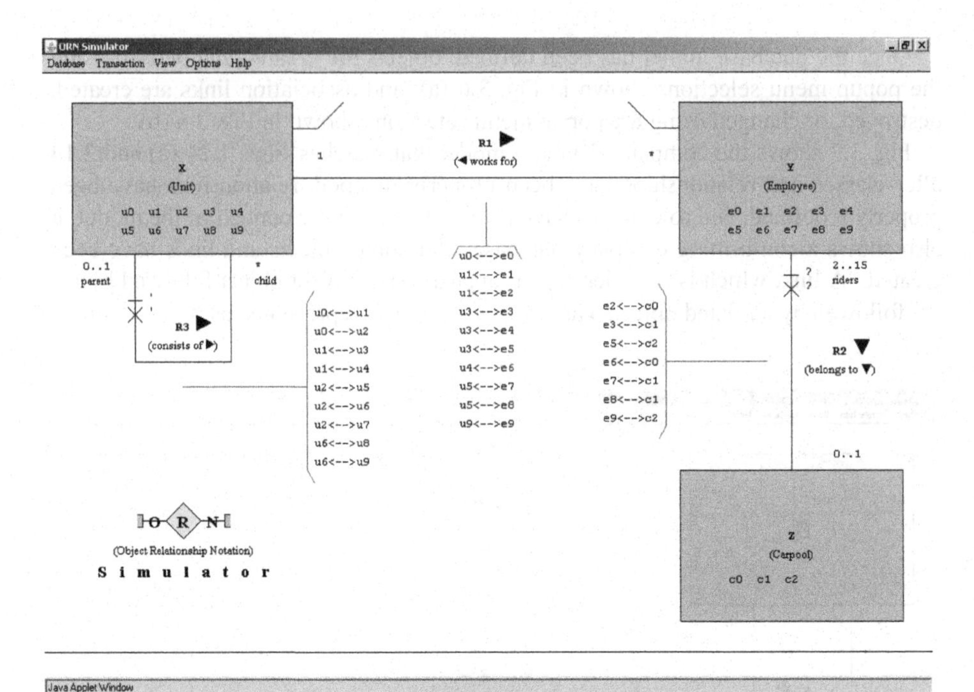

Fig. 3.7 Screen showing initial company database model and database

Fig. 3.8 shows the dropdown menu selections for the menu bar "Database", "Transaction", and "Options" selections. These selections provide the remaining functionality of the ORN Simulator and are briefly explained in Table 3.1. Some additional information about these selections is as follows:

- The dump for the database given in Fig. 3.7—including its meta data, i.e., model—is given in Fig. 3.9. Clicking "Database Restore…" and then copying and pasting this text into the displayed textbox restores (or creates) the database shown in Fig. 3.7.
- "Generate …" selections generate skeletal DDL specifications that can be used to implement the modeled database as a real object or relational database. ODDL is discussed in Chapter 7 and ORN Additive is discussed in Chapter 6. Clicking "Generate T-SQL w/ ORN Additive" for the model shown in Fig. 3.7 results in the screen shown in Fig. 3.10. (Readability indicators were first removed from the mapped association names as these names are used to form constraint names.)
- A Commit was done after creating the objects and links shown in Fig. 3.7. All lower bound multiplicities must be satisfied at the commit of a transaction.
- The "Set RXC mode" and "Reset RXC mode" are discussed in the next section.
- Clicking on a class id, for instance X, then clicking on "Create Some Objects" creates SomeObjectsMax minus the current number of X objects. Selecting "SomeObjectsMax…" under "Options" allows the user to set this parameter to a value between 2 and 10.

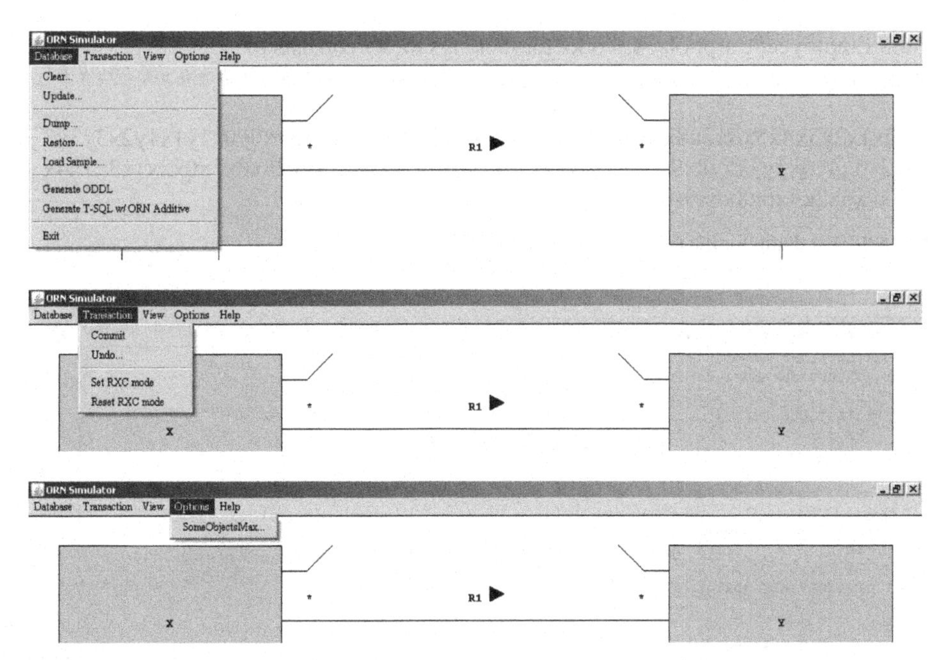

Fig. 3.8 Partial screens showing menu bar selections for Database, Transaction, and Options

Table 3.1 Functions provided by the menu selections shown in Fig. 3.8

Database	
Clear...	Clears all objects and links from the database without changing mappings or relationship types.
Update...	Prompts the user to "Select an object class or relationship for update by clicking on a class id or relationship id, respectively."
Dump...	Dumps the database as text into a textbox so that it can be copied to a text file, saved, and later used to restore the database.
Restore...	Restores the database by using the text that was previously copied and saved. A textbox is provided into which the text is pasted.
Load Sample...	Loads a sample database that can be updated for demonstration or experimentation.
Generate ODDL	Generates partial Object Database Definition Language (ODDL) specifications for the database as it has been modeled.
Generate T-SQL w/ ORN Additive	Generates Microsoft T-SQL with ORN Additive specifications for the database as it has been modeled.
Exit	Exits the ORN Simulator program.
Transaction	
Commit	Commits all changes made to the database since the initiation of the ORN Simulator, the last Database Clear, the last Commit, or the last Undo.
Undo...	Undoes all changes made to the database since the last Commit.
Set RXC mode	Sets Relationship eXChange mode, which suspends the implicit deletions of related objects to facilitate the interchange of existing association links between objects.
Reset RXC mode	Resets Relationship eXChange mode.
Options	
SomeObjectsMax...	Provides a textbox to set SomeObjectsMax, which determines the maximum number of objects created for the Create Some Objects popup menu selection

Unit#Employee#Carpool#◄ works for#belongs to ▼#consists of
►###riders#parent####child##1#2..15#0..1#*#*#0..1#*#*# #?#'# # # # # # #?#'# #
# # #
#/x0x1x2x3x4x5x6x7x8x9#y0y1y2y3y4y5y6y7y8y9#z0z1z2#x0y0x1y1x1y2x3y3x3
y4x3y5x4y6x5y7x5y8x9y9#y2z0y3z1y5z2y6z0y7z1y8z1y9z2#x0x1x0x2x1x3x1x4x
2x5x2x6x2x7x6x8x6x9##W

Fig. 3.9 Text dump for the database shown in Fig. 3.7

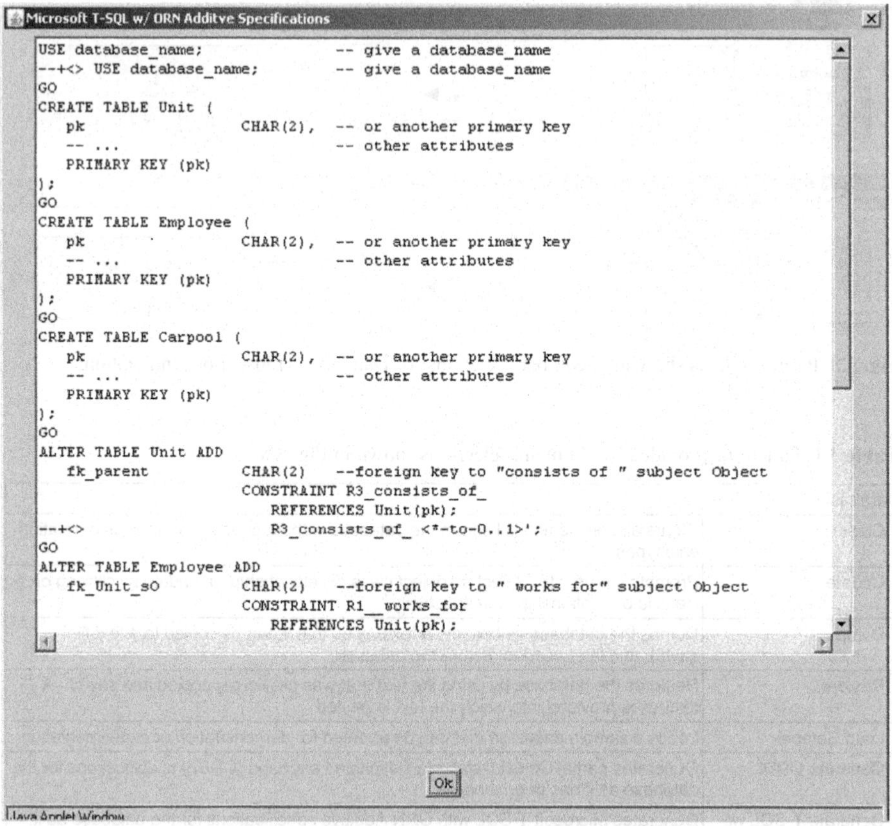

Microsoft T-SQL w/ ORN Additve Specifications

```
USE database_name;                -- give a database_name
--+<> USE database_name;          -- give a database_name
GO
CREATE TABLE Unit (
    pk                  CHAR(2),   -- or another primary key
    -- ...                         -- other attributes
    PRIMARY KEY (pk)
);
GO
CREATE TABLE Employee (
    pk                  CHAR(2),   -- or another primary key
    -- ...                         -- other attributes
    PRIMARY KEY (pk)
);
GO
CREATE TABLE Carpool (
    pk                  CHAR(2),   -- or another primary key
    -- ...                         -- other attributes
    PRIMARY KEY (pk)
);
GO
ALTER TABLE Unit ADD
    fk_parent           CHAR(2)    --foreign key to "consists of " subject Object
                        CONSTRAINT R3_consists_of_
                            REFERENCES Unit(pk);
--+<>                       R3_consists_of_ <*-to-0..1>';
GO
ALTER TABLE Employee ADD
    fk_Unit_sO          CHAR(2)    --foreign key to " works for" subject Object
                        CONSTRAINT R1__works_for
                            REFERENCES Unit(pk);
```

Ok

Java Applet Window

Fig. 3.10 Generated T-SQL w/ ORN Additive for the database shown in Fig. 3.7

3.2 Verifying Association Semantics

Now we experiment with our associations to verify some of their semantics.

Figs. 3.11 through 3.13 shows the sequence of screens that occur in an attempt to delete `u0`. An exception results because this deletion would violate the lower bound multiplicity of 1 for the Unit class in the "works for" association. Here, the implicit default destructibility binding for Unit in the R1, or "works for" association, is operative. The <R1 in the exception message refers to the inverse relationship to R1>. In the <R1 relationship, Employee is the subject class and Unit is the related class.

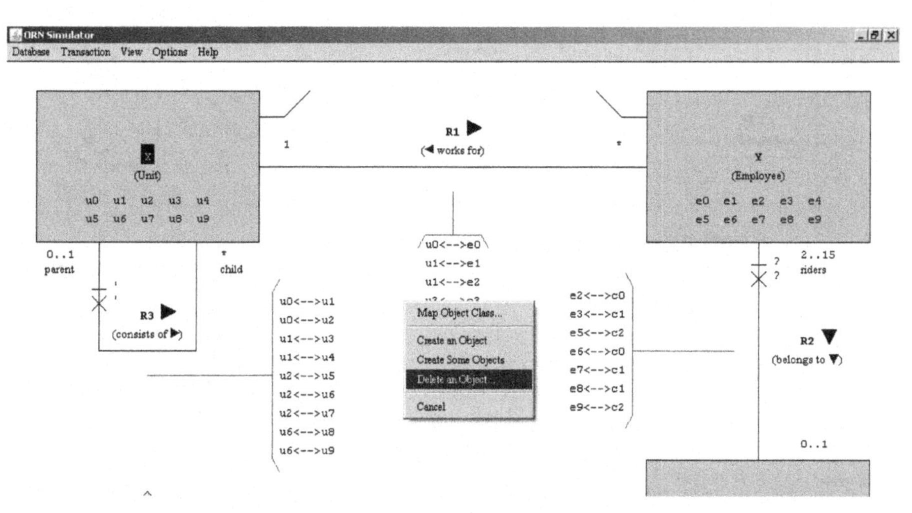

Fig. 3.11 Partial screen after clicking on class X and selecting Delete an Object…

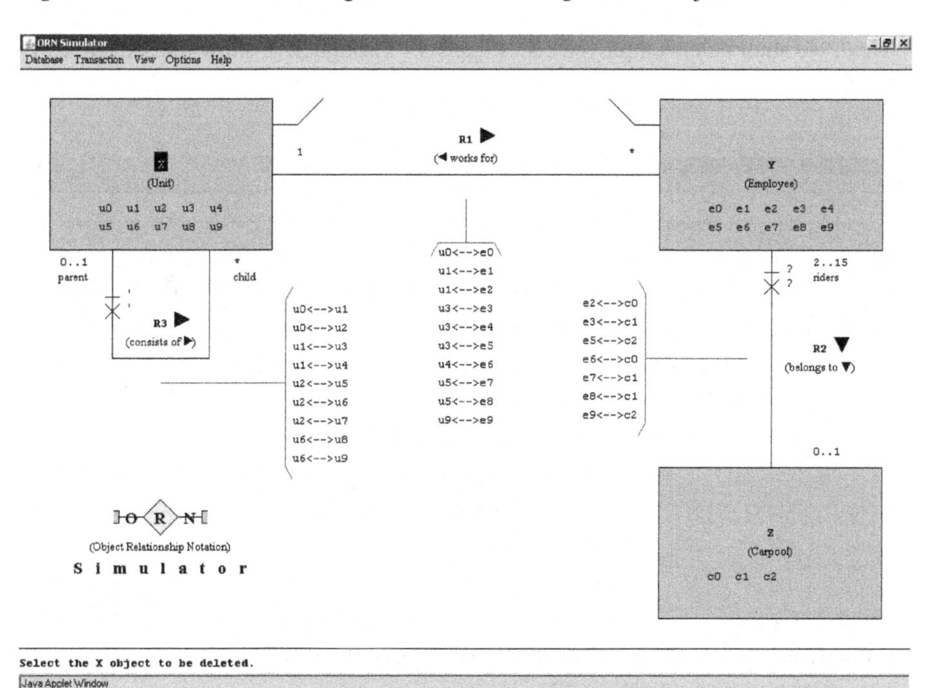

Fig. 3.12 Screen after clicking on Delete an Object…, showing prompt to "Select the X object …"

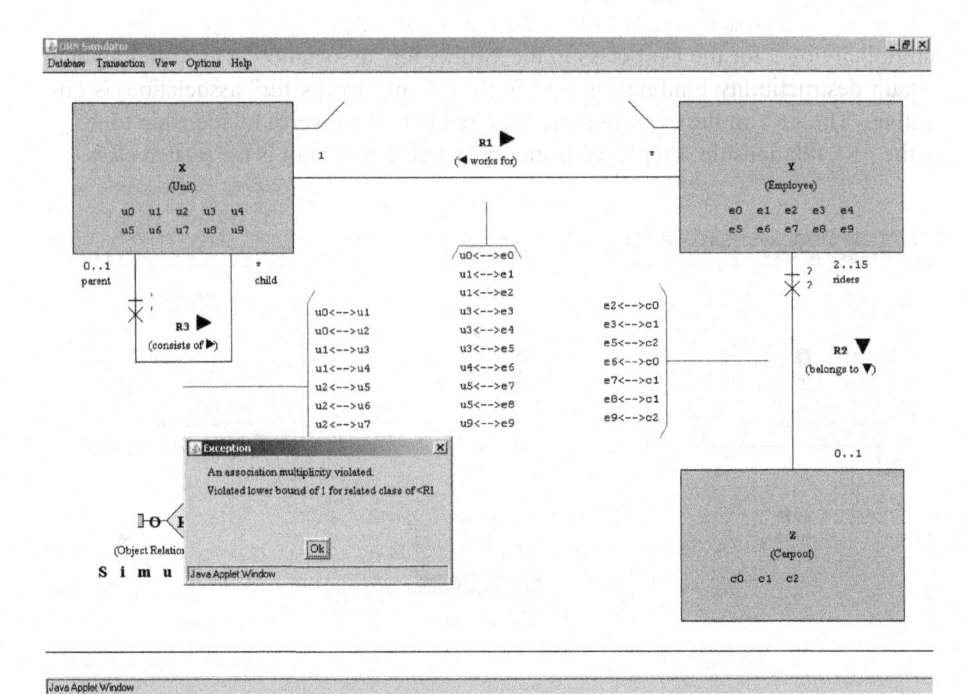

Fig. 3.13 Screen showing exception after clicking on object u0

Next, we shall delete employee e2. Figs. 3.14 through 3.16 show the sequence of screens.

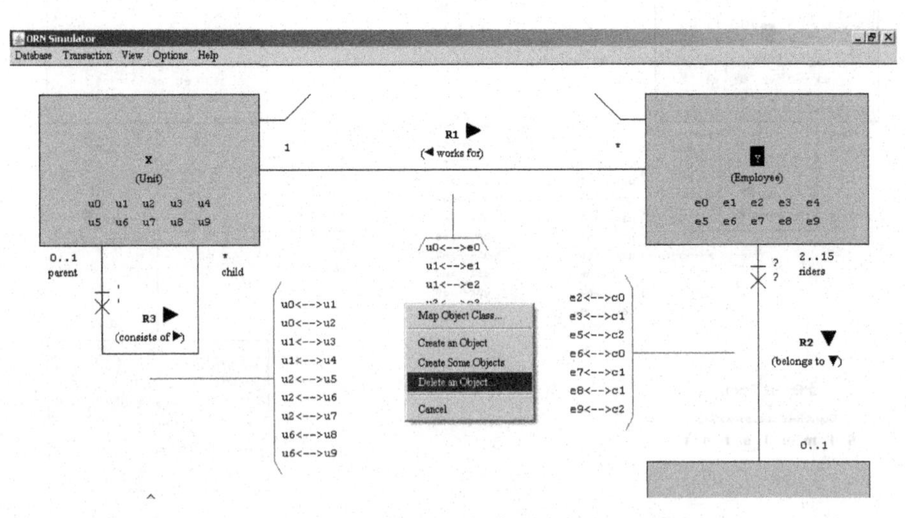

Fig. 3.14 Partial screen after clicking on class Y and selecting Delete an Object…

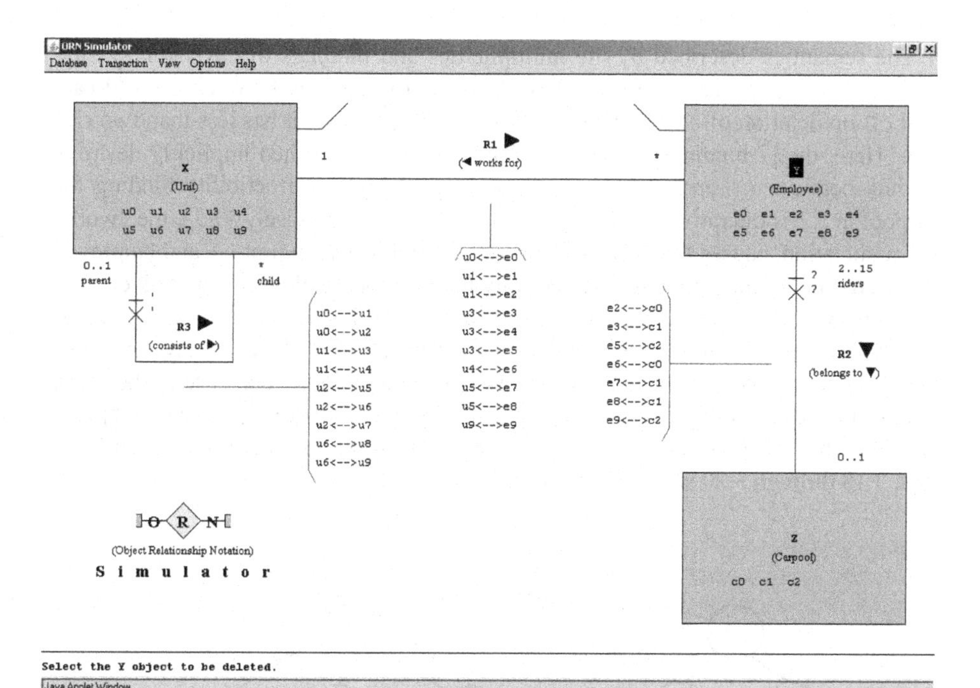

Fig. 3.15 Screen after clicking on Delete an Object..., showing prompt to "Select the Y object ..."

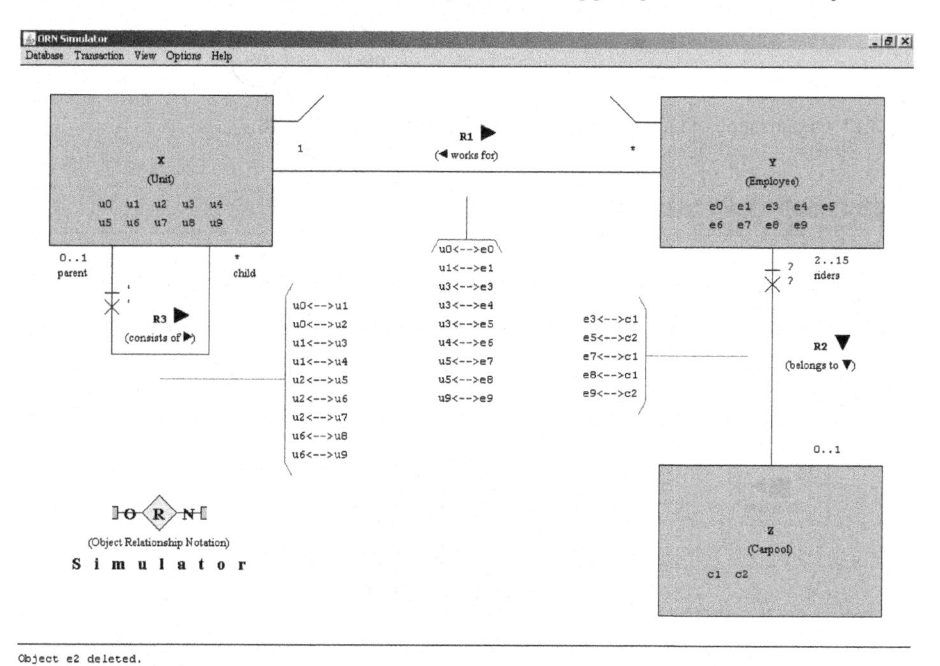

Fig. 3.16 Screen showing results after clicking on object e2

The semantics described by the multiplicities and bindings of ORN for R2, the "belongs to" association, cause the DBMS to destroy the link between e2 and carpool c0 on deletion of e2 and then delete car pool c0 (since it has less than two riders). Here, the |? binding is operative. The deletion of c0 then implicitly destroys the link between e1 and c0. Here, the implicit default destructibility binding for Carpool in R2 is operative. Also, the ORN for R1 (more precisely <R1), the "works for" association, causes the DBMS to destroy the link between unit u1 and employee e2. Here, the implicit default destructibility binding for Employee is operative.

Lastly, we verify the **X'** binding for the parent role in R3, the "consists of" association. This association records the current organization of the company. An "org chart" reflecting the current links between organizations in the database is shown as an object diagram in Fig. 3.17. Units u2, u6, u7, and u8 currently have no employees. We shall remove unit u2 from u0, i.e., destroy the link between u0 and u2. Figs. 3.18 through 3.20 show the sequence of screens.

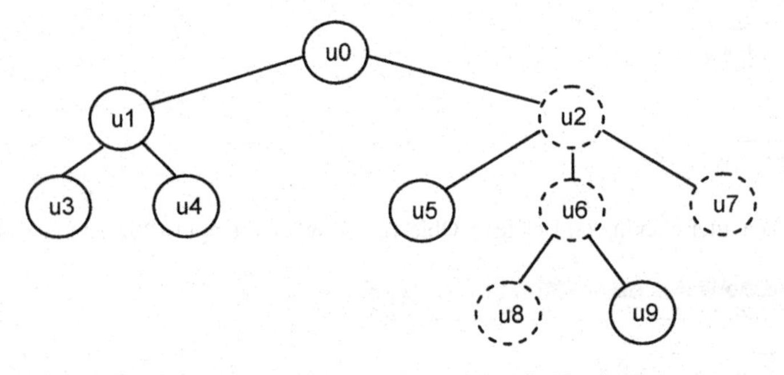

Fig. 3.17 Organization of units as recorded in database shown in Fig. 3.18

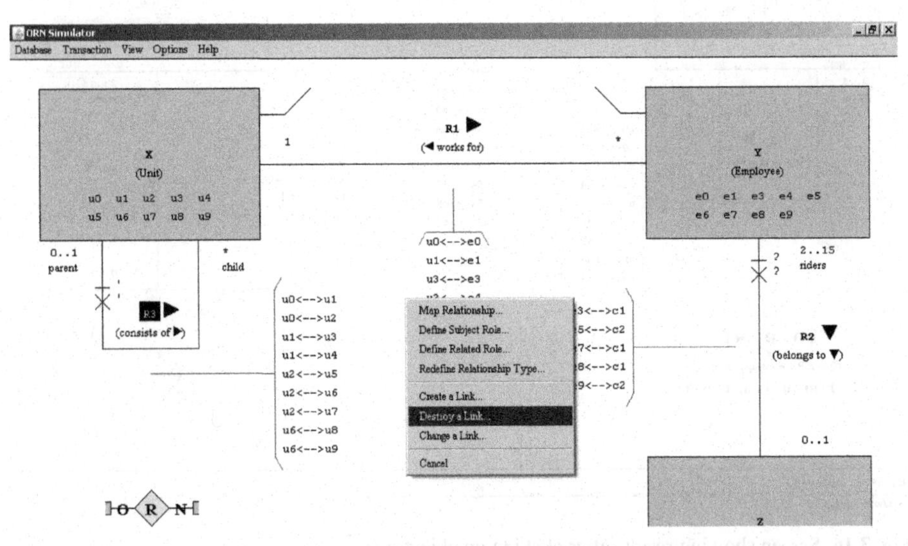

Fig. 3.18 Partial screen after clicking on relationship R3 and selecting Destroy a Link...

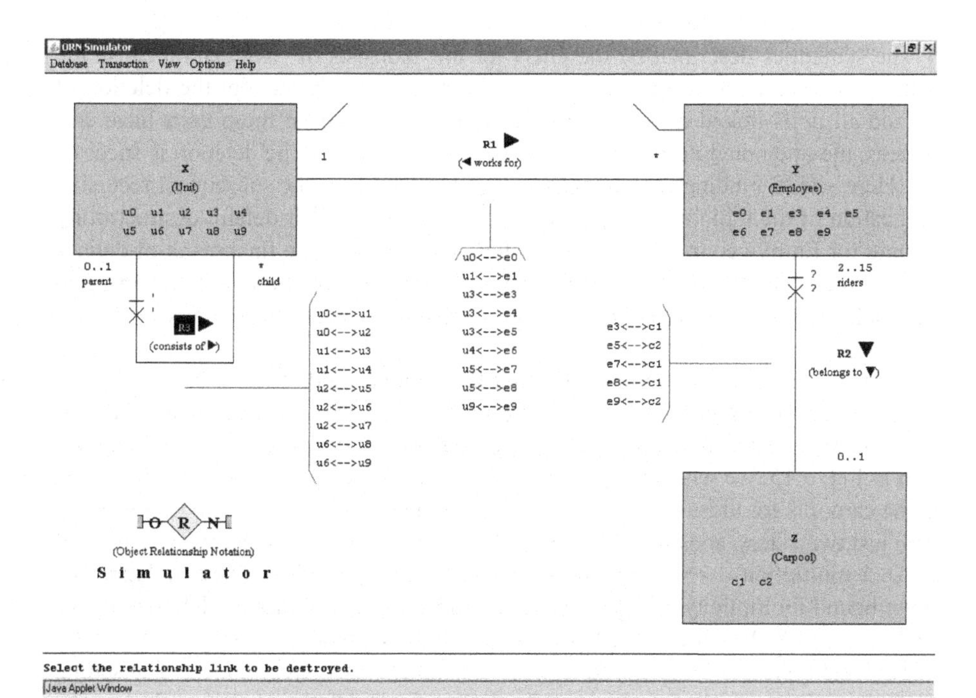

Fig. 3.19 Screen after clicking on Destroy a Link…, showing prompt "Select the relationship …"

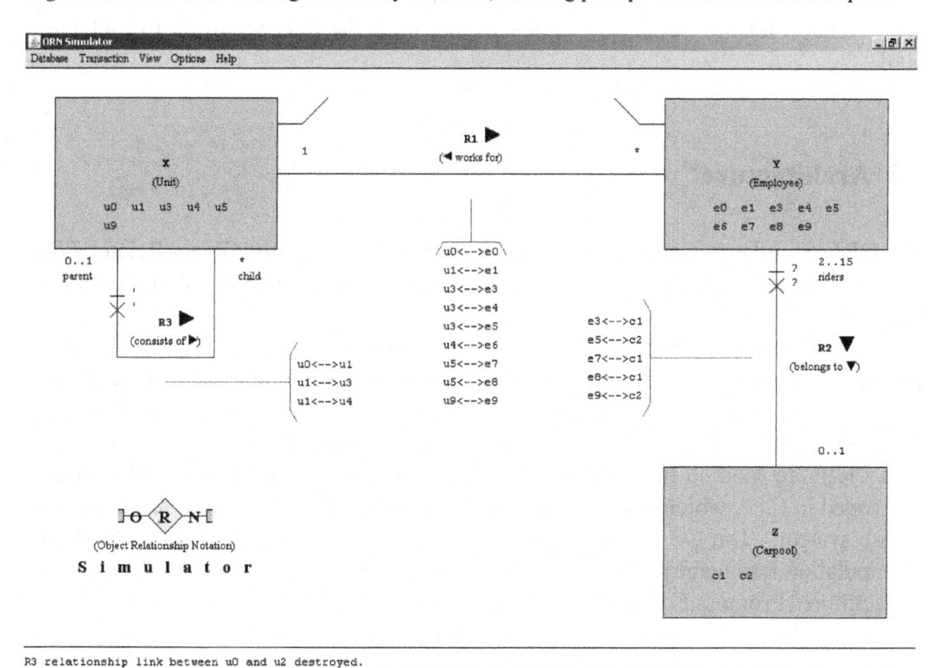

Fig. 3.20 Screen showing results after clicking on link `u0<-->u2`

The semantics described by the ORN for the "consists of" association cause the DBMS on destruction of the u0<-->u2 link to essentially attempt the deletion of u2 and all units linked directly or indirectly to u2. If any of these units have employees, the attempted deletion of the unit is undone. If not, the deletion is successful. Here, the X' binding for parent in R3 is operative, as is the subsequent recursive application of the |' binding for parent in R3, as is the implicit default destructibility binding for Employee in R1, the "works for" association. The final result of delinking u2 from u0 is the removal of u2, u6, u7, and u8, all of the units reporting to u0 that have no employees. Units u5 and u9 remain as independent units, with their employees.

Before concluding this section, the "Set RXC mode" and "Reset RXC mode" selections under Transaction, which are documented in Table 3.1, need more explanation. Suppose that in the original company database, as first seen in Fig. 3.7 and last seen in Fig. 3.15, we wish to exchange the carpools for employees e2 and e9. That is, the carpools for these riders have agreed to trade them. Both c0 and c2 currently have just two riders, and an employee can be in only one carpool at a time because of the 0..1 multiplicity. So how do we "cleanly" do this trade without violating the lower bound multiplicity of 2 for Employee and causing the implicit deletion of carpool c0 or c2? We click "Set RXC mode," destroy links e2<-->c0 and e9<-->c2, create links e2<-->c2 and e9<-->c0, and "Reset RXC mode." The setting of RXC mode suspends the otherwise implicit carpool deletions of c0 and c2. The subsequent transaction commit, which we must do, will ensure that the 2 has not been violated for the carpools, i.e., that we put back the rider we owed to each carpool.

3.3 Architecture

The ORN Simulator is actually an application program that uses Object Relater *Plus* (OR+), a prototype ODBMS (Ehlmann and Riccardi 1997b). OR+ is discussed in Chapter 7.

The tool is implemented with a front-end user interface component and a back-end database management component. The front-end is implemented in Java and runs as an applet client. The back-end is implemented in Java and C++ and runs as an application/database server. Communication between client and server is via Java's Remote Method Interface (RMI). The remote methods are Java native methods coded in C++, which call still other C++ methods defined as part of an object database system. The C++ code manipulates the database using the Object Database Manipulation Language (ODML) of OR+. OR+ implements ORN as an extension to ObjectStore (Progress Software Inc 2006).

3.4 Benefits

The benefits of a tool such as the ORN Simulator for database developers are significant. First, they can more precisely model association semantics. Thus, any inconsistencies in association semantics can be detected. This was not discussed in this chapter but is discussed at length in Chapter 9. Second, association semantics can be verified by simulation before being implemented. This can eliminated wasted effort.

Another significant benefit occurs when the DBMS supports ORN. In the ORN Simulator the user can click on "Database" in the menu bar, then click on one of two "Generate ..." selections. The tool then automatically generates the partial DDL specifications to implement the modeled database as an object database or relational database, depending on the selection. These DDL specifications must of course be processed by a DBMS that automatically maintains the modeled association behavior. Such support for ORN increases data integrity and significantly reduces the amount of database applications code that must be implemented, tested, and maintained by database developers.

The benefits of a tool such as the ORN Simulator for students of software engineering and database management are also significant. They can quickly model and remodel object classes and associations and see firsthand the concrete realizations of their models. This facilitates a more rapid and better understanding of the abstractions inherent in the data model notations. The tool also facilitates student understanding of the concepts involving transactions in database management. It forces students to either commit or abort their changes. When creating objects, they learn constraint checking can be deferred (for lower bound multiplicities). When creating links, they learn constraint checking can also be immediate (for upper bound multiplicities). And finally, when committing their transactions, they quickly discover that all database constraints must be satisfied on transaction commit.

Chapter 4

Association Patterns

Emerging from a Variety of Association Types

When combined with multiplicities, the bindings of ORN allow the semantics of a large variety of association types to be specified. As many of these types are modeled in numerous application domains using ORN, some patterns begin to emerge. Each such pattern can be described as an *association pattern.*

Association patterns provide guidance for modeling the associations that occur among objects in the real world when these objects are implemented in computer applications. The patterns help the designer better understand and more precisely define the semantics of these associations, which allows them to be more easily and properly implemented. This chapter describes a number of association patterns using ORN and by doing so provides evidence for the effectiveness of this notation. It then shows how the development of database systems can be improved by an approach that builds a database model using association patterns and then implements the model by mapping it into an ORN-extended database definition that is supported by a DBMS. The feasibility of this approach and the applicability of our association patterns have been validated by DBMS research prototypes and by the modeling, implementing, and testing of numerous associations.

This chapter is organized into four sections. Section 4.1 can be skipped by the more practical minded. It provides context for association patterns—explaining what they are, what they are not, and what they are like as well as how they relate to analysis patterns, design patterns, patterns in general, and ORN. Section 4.2 is the main focus of the chapter. It first explains how association patterns are to be described and then describes seven basic patterns and fourteen patterns in all as three of the basic patterns have major variations. Section 4.3 shows how association patterns can be used to more productively model and implement database systems. Section 4.4 provides examples of association types that do not conform to any defined association pattern. Section 4.5 concludes by summarizing the motivations for association patterns and their significance.

4.1 Context

The association patterns that are the subject of this chapter are neither the "association patterns" that are discovered in data mining, e.g. (Tan et al. 2006), nor the "association pattern" expressions of an object-oriented database language, e.g. (Guo et al. 1991). Instead, they are patterns for the types of associations relevant to data

B.K. Ehlmann, *Object Relationship Notation (ORN) for Database Applications*,
Advances in Database Systems 39, DOI 10.1007/978-0-387-09554-7_4,

modeling. Three such *association patterns* were introduced by Fowler (1997); however, since then (as far as this author knows), no similar patterns have been proposed.

The closest things to such patterns are the proposed characteristics and limited notations for identifying the semantics of certain types of relationships, which are discussed further in Chapter 5. These are the weak entity in the ER model (Chen 1976), generalization/specialization in semantic and object models (Smith and Smith 1977), the characteristics ascribed to whole-part relationships—e.g., (Albert et al. 2003), (Barbier et al. 2003), (Odell 1994), (Winston et al. 1987)—and the shared aggregation and composition that are given special notation and semantics in the Unified Modeling Language (UML) (OMG 2005). The patterns described here, however, rather than focusing on certain types of relationship—like whole-part relationships or those with weak entities—address associations in general.

In purpose, association patterns are similar to *design patterns* (Gamma et al. 1995) and *data model patterns* (Hay 1996), also termed *analysis patterns* (Fowler 1997). They differ, however, from these patterns in level and focus.

Design patterns are "descriptions of communicating objects and classes that are customized to solve a general design problem in a particular context" (Gamma et al. 1995). They are normally used during systems design and generically describe the classes, relationships, methods, and constraints that provide a solution to a common design problem in developing software. An example of a design pattern is the Adapter pattern (Gamma et al. 1995), which provides for the conversion of one interface of a class into another.

Analysis patterns are "groups of concepts that represent a common construction in business modeling" (Fowler 1997). They are normally used during requirements analysis and database conceptual modeling. Analysis patterns generically describe application-related classes, relationships, and constraints for modeling similar concepts and requirements that are common to many application domains. An example of an analysis pattern is the Party pattern (Fowler 1997), which provides for the conceptual representation of people and organizations common to many applications.

Association patterns are descriptions of the semantics that define common types of associations, i.e., structural relationships, that occur between objects. Like analysis patterns, they are normally used during the requirements analysis and conceptual modeling phases of software development. An association pattern, as defined in this paper, generically describes the multiplicities and *referential actions* for defining an **application-independent** association type that is common to many application or solution domains. The pattern generically names the association type and the participating classes to indicate the roles that related classes and objects play in the association. The focus of association patterns is more limited than that of design and analysis patterns. As such, they can be used at a lower level to better define the associations within these latter patterns. An example of an association pattern is the "is an instance of" pattern, which relates the concrete objects of a "Concrete Class" to the abstract objects of an "Abstract Class."

The purpose of an association pattern is the same as that of any pattern, which is to increase the productivity of design by turning what would have seemed like a unique problem for the novice into an ordinary problem, one like many similar problems often encountered by the expert.

The referential actions that are mentioned above and used to describe association patterns are the bindings of ORN. As indicated in Chapter 1 and as will be more fully discussed in Chapter 5, ORN bindings provide actions more varied than the referential actions of relational DBMSs. These actions can be based on multiplicities and given at a conceptual level, i.e., in ER (Chen 1976) or UML Class Diagrams (OMG 2005, Ehlmann and Yu 2002). They implement a variety of association semantics and define the extent of complex objects. The bindings of ORN are fundamental to describing association patterns, which to some extent provides a validation for their inclusion within ORN and their semantics.

4.2 Pattern Descriptions

This section describes seven basic association patterns. Actually, fourteen different patterns are described since three of the basic association patterns have major variations with slightly different semantics. Table 4.1 lists these patterns and variations along with the subsection in which they are described. Although these patterns are relevant to a great number of the associations seen in the data models for specific applications, no claim is made that they are exhaustive. Also, of course, not all associations can be classified as fitting into any kind of pattern. Examples of such associations are given in Section 4.4.

Table 4.1 Summary of Association Patterns and Their Major Variations

Pattern or Major Variation	Section
"is defined by"	4.2.1
"is recorded for"	4.2.2
Case 1 (Exclusive Use)	4.2.2.1
Case 2 (Shared Use, Single Association)	4.2.2.2
Case 3 (Shared Use, Multiple Associations)	4.2.2.3
"is a realization of"	4.2.3
"is associated by"	4.2.4
"is an update of"	4.2.5
"is a part of"	4.2.6
Case 1 (Shared Aggregation, Independent Existence)	4.2.6.1
Case 2a (Shared Aggregation, Dependent Existence, Single Association)	4.2.6.2
Case 2b (Shared Aggregation, Dependent Existence, Multiple Associations)	4.2.6.3
Case 3a (Composite Aggregation, Independent Existence)	4.2.6.4
Case 3b (Composite Aggregation Dependent Existence)	4.2.6.5
"is a"	4.2.7
Case 1 (Optional Subclasses)	4.2.7.1
Case 2 (Mandatory Subclasses)	4.2.7.2

Each association pattern is described by providing the following elements in the order given below:

- an identifying generic name for the basic pattern and type of association, e.g., "is a part of", and when any major variations exist on the basic association type, an identifying case number along with some descriptive terms , e.g., "Case 1 (Shared Aggregation, Independent Existence)";
- generic names for each of the associated classes, a description of the association type's distinguishing semantics, a partial ORN (or <association>) that reflects the generally desired semantics of the association type, and a model of the association type using an ORN-extended UML class diagram;
- an *Examples* segment providing three examples of associations that fit the pattern, at least for some application, and the UML model for one of these examples;
- a possible *Variations* segment explaining minor variations on the pattern's <association> that give slightly different, but perhaps more desirable, semantics; and finally,
- a possible *Comments* segment providing significant observations on the pattern, e.g., its dependence on a specific binding.

Each association pattern or major variation begins at or near the top of a new page.

In describing an association pattern and its ORN, some conventions have been adopted. A ... denotes a portion of the <association> not prescribed by the pattern and thus free to be given as desired. An m denotes a multiplicity. m_L is the lower bound, and m_U is the upper bound. Thus $m = m_L..m_U$ where $m_L \leq m_U$ or $m_U = *$. A * multiplicity may be given as 0..* to stress a lower bound of 0. A ! binding is always used with a 1 multiplicity when a ? binding would also work (but would mean a lower bound check). Finally, for a class C, an object of type C may be referred to as a "C object," "a C," or "a c."

All of the bindings of ORN are necessary to accurately define the semantics of the fourteen association patterns. Each type of binding is explained within a **Review** "sidebar" in the *Examples* segment of a pattern description when the type of binding is first encountered. The explanation is similar to that given in Chapter 2 for the "works for" association between employees and units but is done in the context of the given association example that is modeled.

4.2.1 "is defined by" pattern

A *defined object* "is defined by" its association with a minimum number of *defining objects*. If the number of defining objects falls below this minimum, the defined object should be deleted. The ORN for this pattern is …-to-*m*>~? where $m_L \geq 1$. Its model is shown in Fig. 4.1(a).

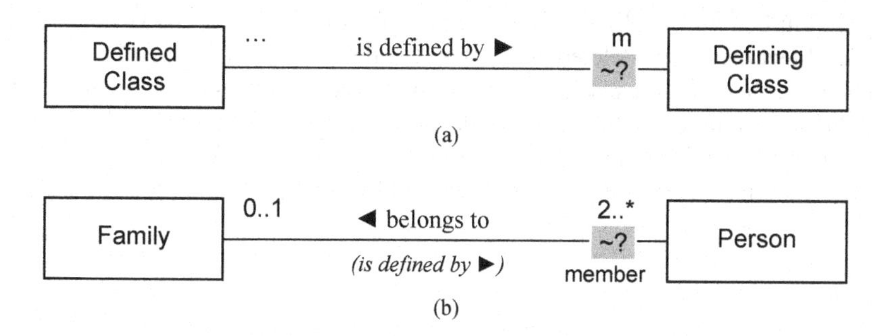

Fig. 4.1 The "is defined by" pattern—(a) generic model and (b) example

Examples. A Family is defined by at least two members, i.e., two Person objects, <0..1-to-2..*>~? (see Fig. 4.1(b)). Inversely, a person belongs to zero or one family. If the number of members in a family falls below two, the family is deleted. (The optional ~ symbol meaning "cascade" is not used in the remainder of the chapter.)

Review of the ? and Default Binding

The ? binding for the Person class implies both a |? and X? binding. The |? binding applies when a Person object is deleted and means (paraphrasing from Table 2.1): a "belongs to" link can be destroyed, but if this violates the multiplicity 2..*, the destruction must be cascaded to the related Family object, i.e., this object must be implicitly deleted. This multiplicity will be violated when the person being deleted is one of just two members belonging to the family.

The X? binding applies when a "belongs to" link is explicitly destroyed and means (paraphrasing from Table 2.1): the link can be destroyed, but if this violates the multiplicity 2..*, the destruction must be cascaded to the related Family object. Again, the multiplicity will be violated when the person being "de-linked," or removed, from the family is one of just two members in the family.

Implicit default binding for the Family class applies when a Family object is deleted and means (paraphrasing from Table 2.1): a "belongs to" link can be destroyed provided this does not violate the 0..1 multiplicity. The 0..1 will not be violated. Thus a family can be deleted despite having persons that belong to it. (The "family" simply dissolves.)

Explicit default binding for Family applies when a "belongs to" link is explicitly destroyed and means (paraphrasing from Table 2.1): the link can be destroyed provided this does not violate the 0..1 multiplicity. Again, the 0..1 will not be violated.

Thus based on the Family binding, persons can be removed from the family (though as we have seen, based on the Person binding, this could mean the death of the family).

A Polygon is defined by at least three Line objects, <*-to-3..*>?. A Track through a particle detector is defined by at least two connecting Subtrack objects in adjacent layers of the detector, which has eight such layers, <0..1-to-2..8>?.

Variations. An alternative ORN that may be desirable is ...-to-*m*> where $m_L \geq 1$. Here, the implicit and explicit default bindings for the Defining Class mean that a defining object cannot be deleted and a link to a defining object cannot be explicitly destroyed if either would cause the m multiplicity (actually m_L) to be violated. This forces the defined object to be explicitly deleted if it is no longer valid.

Comments. The ? binding is necessary for the semantics of this pattern when automatic deletion of invalid defined objects is desired. (A ! binding can be used if $m_U = m_L$.)

4.2.2 "is recorded for" pattern

A *supporting object* "is recorded for" one or more *primary objects*. It may exist independent of any primary object in the real world, but in the system being modeled a supporting object is recorded only because of its relevance to some primary object, and if it is no longer relevant to any primary object, it should be deleted.

4.2.2.1 Case 1 (Exclusive Use).

A supporting object is recorded for only one primary object. The Supporting Class is what is often called a *weak entity type* in ER modeling. The ORN for this pattern is ...-to-1>!, and its model is shown in Fig. 4.2(a).

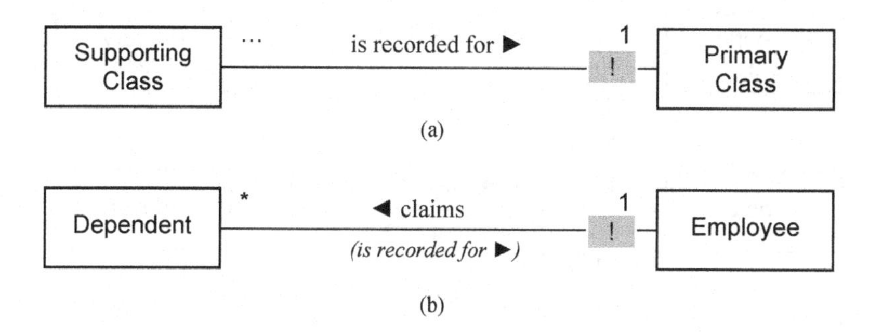

Fig. 4.2 The "is recorded for" pattern, Case 1—(a) generic model and (b) example

Examples. A Dependent is recorded for an Employee, <*-to-1>! (see Fig. 4.2(b)). A dependent is recorded only because of its relevance to an employee. If a dependent is no longer relevant to an employee, either because its relationship to the employee has been severed or because the employee has been deleted, then the dependent is deleted.

Review of the ! Binding

The |! binding, implied by the ! binding, applies when an Employee object is deleted and means (paraphrasing from Table 2.1): a "claims" link can be destroyed, but the destruction must be cascaded to the related Dependent object. Thus, when an employee is deleted all related dependents will be deleted. The X! binding, also implied by the ! binding, applies when a "claims" link is explicitly destroyed and means (again paraphrasing from Table 2.1): the link can be destroyed, but the destruction must be cascaded to the related Dependent object. Thus, explicitly destroying the relationship between an employee and a dependent will automatically delete the dependent.

An Application is recorded for a JobCandidate, <0..1-to-1>!. An Assessment—e.g., exam, assignment, final grade, etc.—is recorded for a Class, <*-to-1>!.

Comments. The ! binding is sufficient for this pattern but not necessary. The ? binding could have been used since the lower bound multiplicity for the Primary Class is 1. Later we will see a pattern where the ! binding is necessary. The |! aspect of the ! binding is similar to the ON DELETE CASCADE referential action in SQL; however, unlike this action, the |! binding can be used for either end or both ends of a binary association.

4.2.2.2 Case 2 (Shared Use, Single Association)

A supporting object can be recorded for multiple primary objects being related to these objects via just one association. The ORN for this pattern is ...-to-1..*>?, and its model is shown in Fig. 4.3(a).

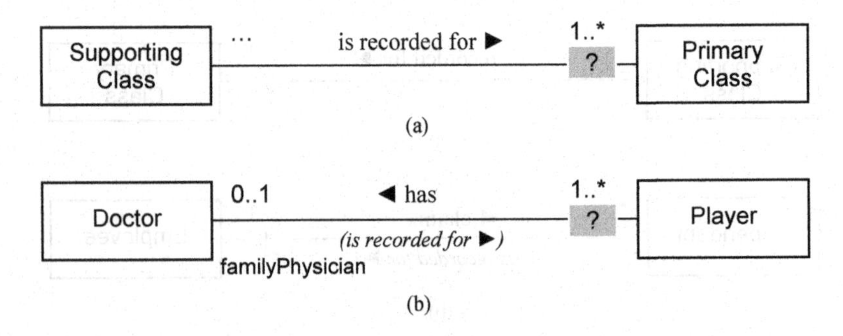

Fig. 4.3 The "is recorded for" pattern, Case 2—(a) generic model and (b) example

Examples. A Doctor is recorded for a soccer league Player, playing the role of a familyPhysician, <0..1-to-1..*>? (see Fig. 4.3(b)). If the last player related to a particular doctor is deleted, the doctor is implicitly deleted. Also, the doctor is implicitly deleted when a doctor no longer serves any player because a player's relationship to the doctor has been severed (via link destruction) or the player has changed doctors (via a link change).

A code Library is recorded for a software system Executable if the executable uses any components in the library, <*-to-1..*>?. If a library is no longer used by any executable, it should not be recorded. For purposes of automatic payroll deduction, a Charity (actually, up to three) is recorded for an Employee, <0..3-to-1..*>?.

Variations. Though rare, an upper bound ≥ 2 can be given for the Primary Class, restricting the reuse of a supporting class object.

Comments. The ? binding is necessary to describe the semantics of this pattern.

This pattern is similar to the "is defined by" pattern. This is because the Supporting Class object's existence depends on being related to at least one Primary Class object. With this pattern, however, a Primary Class multiplicity with lower bound other than 1 is senseless and with upper bound other than * is quite unusual.

4.2.2.3 Case 3 (Shared Use, Multiple Associations)

A supporting object can be recorded for multiple primary objects of the same type or different types, via multiple "is recorded for" associations. For each such association$_i$ with a Primary Class $_i$, where $i = 1, 2, \ldots, n$ and $n \geq 2$, the ORN is *b*<*m*-to-0..*>' where the binding *b* includes a |- if $m_L = 0$. Primary Class $_j$ may be the same class as Primary Class $_k$ where $1 \leq j < k \leq n$. The model for the pattern is shown in Fig. 4.4(a).

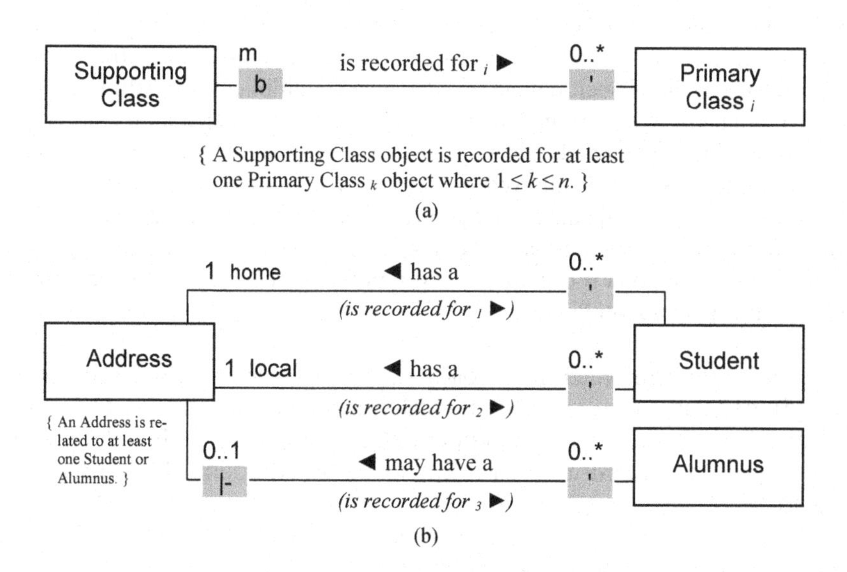

Fig. 4.4 The "is recorded for" pattern, Case 3—(a) generic model and (b) example

Examples. A home Address must be recorded for a Student, <1-to-*>', as well as a local Address, <1-to-*>', though they are often the same (see Fig. 4.4(b)). If known, an Address is also recorded for an Alumnus, |-<0..1-to-*>', perhaps a parent of a student. For many college mailings, only one copy is sent to each address. An address is deleted if it no longer is relevant to any student or alumnus.

Review of the ', |-, and Default Bindings and Link Change

The ' binding for the Student class in the first "has a" association implies both a |' and X' binding. The |' binding applies when a Student object is deleted and means (paraphrasing from Table 2.1): a "has a" link can be destroyed, but an attempt must be made to cascade the destruction to the related Address object; however, this implicit Address object deletion must be undone if it fails, but is required if and only if its undoing would violate the 0..* multiplicity. In this case the 0..* is never violated. Also, in this case the attempted deletion of the related Address object will succeed unless it is also related to another Student or Alumnus object.

This is true because the bindings for the Address class in all of the associations do not permit the implicit destruction of an **existing** link on deletion of an Address object. Each of the implicit default bindings for the "has a" associations means (paraphrasing from Table 2.1): a link can be destroyed provided this does not violate the 1 multiplicity. This 1, however, will **always** be violated. The |- binding for Address in the "may have a" association means (from Table 2.1): a link cannot be destroyed.

The X' binding, implied by the ' binding for all associations, applies to the explicit destruction of a link. Its meaning is similar to that of the |' binding. Essentially, on explicit destruction of a link, the related Address object is implicitly deleted unless it has another link with a Student or Alumnus object.

For explicit link destruction, however, the bindings at both ends of an association are operative. In fact, for each "has a" association, explicit link destruction is not allowed because of the explicit default binding and 1 multiplicity for Address, which would be violated. A link change, however, done as a single operation that replaces the Address object with another is not treated as an explicit link destruction relative to the Address class and is allowed in this case (see footnote 1 in Table 2.1). In such a link change, the old address is deleted if no longer used (based on the X' binding for Student, the related class in the link change). An example of a link change done as a single operation is changing a non-null foreign key value in a relational database to a different non-null value.

A Beneficiary is recorded for a Policy via two associations, playing the role of primaryBeneficiary in one and contingentBeneficiary in the other, <1-to-*>' and |-<0..1-to-*>', respectively. Two EmergencyContact objects, each with name and phone numbers, are recorded for each Employee and Student. Both associations are <2-to-*>'.

Variations. Though rare, an upper bound can be given for the Primary Class, restricting the reuse of a supporting class object.

Comments. The ' binding is necessary for the semantics of this pattern. The |-binding is needed when the lower bound multiplicity for the Supporting Class is 0.

For this particular pattern variation, ORN does not constrain independent supporting objects from being created, thus the need for the constraint given in braces in the model of Fig. 4.4 (b).

If the bindings and multiplicities for the Supporting Class are the same for all associations, a Case 2 pattern can often be used instead of a Case 3 pattern by creating a super class for all Primary Class $_i$ and associating the Supporting Class to the super class.

4.2.3 *"is a realization of" pattern*

A *concrete object* "is a realization of" an *abstract object.* An abstract object should not be deleted if there are any related concrete objects. Once a concrete object is created, its relationship to its abstract object must not be changed or destroyed. The ORN for this pattern is X-<...--to-1>X-, and its model is shown in Fig. 4.5(a).

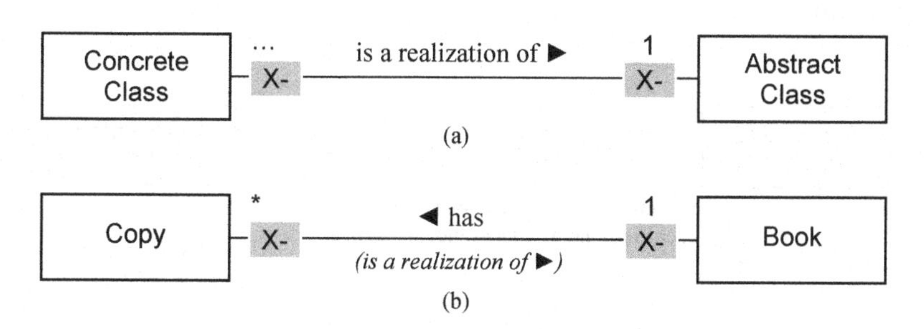

Fig. 4.5 The "is a realization of" pattern—(a) generic model and (b) example

Examples. A Copy is a realization of a Book, X-<*-to-1>X- (see Fig. 4.5(b)). A Book has attributes like ISBN number, title, author, and publisher. A Copy has attributes like bar code number, location, and condition. A book should not be deleted if there are still copies remaining in stock, which is expressed by the 1 multiplicity and the implicit default binding for Book in the association. Also, once a copy is created, its relationship to a book must not be changed or destroyed.

Review of the X- Binding and Link Change

The X- binding for Copy applies to the explicit destruction of a "has" link and means (according to Table 2.1): the link cannot be destroyed. Analogous meaning applies to the X- binding for Student. Both X- bindings result in all link changes being disallowed.

A Class (or Section) is a realization of a Course, X-<*-to-1>X-. An ActualFlight is a realization of a ScheduledFlight, again X-<*-to-1>X-.

Variations. A slightly revised ORN, ...X-<...-to-1>|!X-, may be desirable when the "is a realization of" relates objects within planning or design applications. Unlike applications that store data representing real world objects, here allowing the deletion of an abstract object when related "concrete" objects exist is often desirable. The |! means that the related concrete objects will be implicitly deleted. For example, a ClassUsage, which may appear in many different models in a CASE tool, is a realization of a ClassDefinition, which is abstractly defined at a system level, X-<*-to-1>|!X-. To delete all usages of a class from all models, the user need only delete the related ClassDefinition.

Comments. The X- binding is necessary for the semantics of this association pattern.

4.2.4 "is associated by" pattern

An *associated object* "is associated by" an *associating object* to one or more other associated objects representing a many-to-many or *n*-ary association. For each such association$_i$ between an Associated Class $_i$ and the Associating Class, where $i = 1, 2, \ldots, n$ and $n \geq 2$, the ORN is !<1-to-... . Associated Class $_j$ may be the same class as Associated Class $_k$ where $1 \leq j < k \leq n$. The model for the pattern is shown in Fig. 4.6(a).

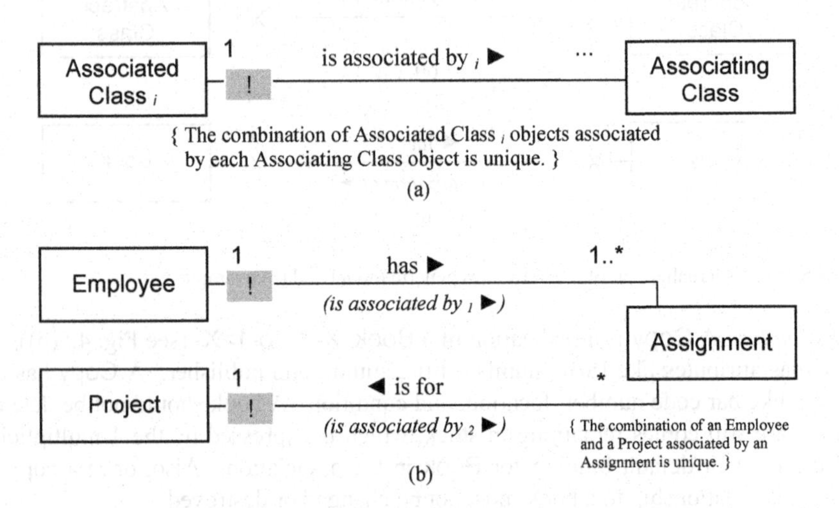

Fig. 4.6 The "is associated by" pattern—(a) generic model and (b) example

Examples. An Employee and a Project are associated by an Assignment (see Fig. 4.6(b)). The association between Employee and Assignment is !<1-to-1..*>, since an employee must be assigned to at least one project, and the association between Project and Assignment is !<1-to-*>. If an employee or a project is deleted, all related assignments are implicitly deleted; however, in deleting a project, a lower bound violation occurs if any related assignment is the only one for an employee.

In a graph two Node objects are associated by an Adjacency object, containing the weight of the connecting edge. Both associations between Node and Adjacency are !<1-to-*>. A Part, a Project, and a Vendor are associated by a Procurement, representing a ternary, many-to-many-to-many association. The associations between each of the associated classes—Part, Project, and Vendor—and Procurement are !<1-to-*>.

Variations. Some revisions to the normal ORN may be desirable. A <1-to-... disallows the deletion of an Associated Class object, e.g., an employee, if any related Associating Class objects, e.g., an assignment, exist (since the 1 multiplicity for the Associated Class would be violated). A |!X-<1-to-...>X- or a X-<1-to-...>X- disallows any explicit link destructions (or changes). The ability to destroy links is normally unimportant for this type of association (and perhaps not recommended) since the insertion and deletion of Associating Class objects, e.g., assignments, essentially creates and destroys relationships between Associated Class objects, e.g., between employees and projects.

4.2.5 "is an update of" pattern

A *versioned object* "is an update of" another such object, i.e., a newer object is an updated version of an older object. Only the first version of an object is not an updated version of an older object. If an object is deleted, all updated versions should also be deleted. The link between an updated version of an object and its old version cannot be destroyed; however, a new version of the object can replace an updated version, in which case the original updated version is implicitly deleted. The ORN for the association pattern is X-<0..m_U-to-0..1>! where $m_U \geq 1$. Its model is shown in Fig. 4.7(a).

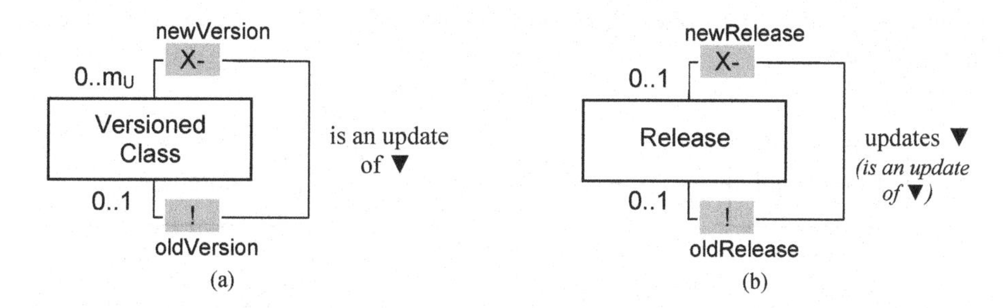

Fig. 4.7 The "is an update of" pattern—(a) generic model and (b) example

Examples. A software Release object is an update of another such Release object, X-<0..1-to-0...1>!, representing a newer release (see Fig. 4.7(b)). The Release class includes attributes like releaseNum and descOfChanges. The link between an updated release and its predecessor cannot be severed. It can, however, be changed to replace the updated release by another if done by a single, atomic operation. The original updated release is implicitly deleted.

A BluePrint is an update of an older BluePrint, X-<*-to-0..1>!. A textbook Edition is an update of an earlier textbook Edition, X-<0..1-to-0..1>!.

The Versioned Class often has an "is a realization of" association with a Configuration Item Class, which invariably describes the item placed under configuration control management. For example, a software Release "is a realization of" a SoftwareItem, the latter having attributes itemName and latestRelease. (A software Copy is a (even more concrete) realization of a software Release).

Comments. The ! and X- bindings are necessary for the semantics of this pattern.

4.2.6 *"is a part of" pattern*

A component object "is a part of" an aggregate object. The semantics of this basic pattern are quite varied and are described in a manner consistent with the aggregate associations as defined in UML (OMG 2005). The semantics include the transitive and antisymmetric properties for relationships and a notion of a "whole" and a "part" in a *whole-part relationship* (Barbier et al. 2003, Winston et al. 1987). The "is a part of" may also be recursive, meaning the component object can itself be an aggregate object and are either the same object type or the component object type "is a" object type that can also be an aggregate object type.

4.2.6.1 Case 1 (Shared Aggregation, Independent Existence)

The component object can be part of multiple aggregate objects of the same type or different types, i.e., it can be shared, and can also exist separate from, i.e., independent of, an aggregate object. The ORN for the association pattern is …-to-0..m_U> where $m_U \geq 2$ if all aggregates are of the same type, else $m_U \geq 1$. Its model is shown in Fig. 4.8(a).

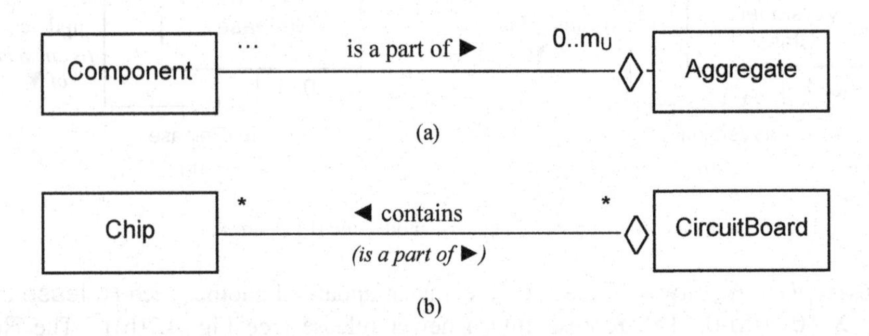

Fig. 4.8 The "is a part of" pattern, Case 1—(a) generic model and (b) example

Examples. An object representing a type of Chip can be part of an object representing a type of CircuitBoard, <*-to-*> (see Fig. 4.8(b)). (Note that an actual chip is a realization of a Chip.) A television Channel is part of a Package offered by a cable company, <*-to-*>. A Line can be part of a number of Polygon objects, ?<3..*-to-*>.

Comments. The last example shows how two patterns can be used for the same association, one for each end. Here, a polygon "is defined by" three or more lines.

4.2.6.2 Case 2a (Shared Aggregation, Dependent Existence, Single Association)

A component object can be part of multiple aggregate objects of the same type, via a single association, but cannot exist separate from, i.e., independent of, an aggregate object.

If the association is not recursive, the ORN for the association pattern is ...-to-1..*>?, and the model is shown in Fig. 4.9(a).

If the association is recursive, the ORN for the pattern is *b*<*m*-to-0..*>' where *b* includes |- if m_L = 0. The model is shown in Fig. 4.10(a). For a recursive association, the lower bound on the aggregate end is 0 since not all objects can play the role of a component object, i.e., there must be at least one aggregate that is not a component. (The 0 allows an object to be created that is not part of another object. Such an object must be an aggregate and not an independent component.)

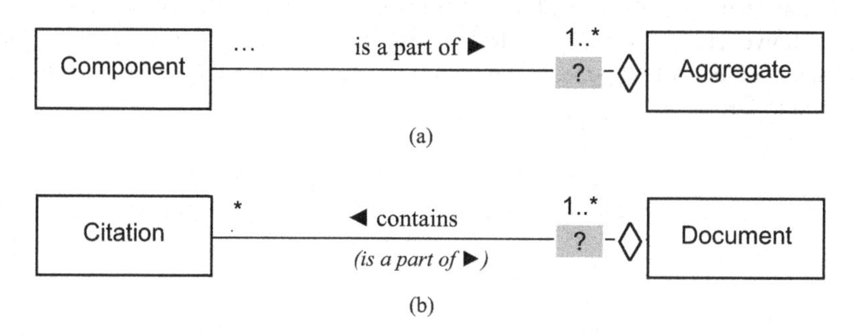

Fig. 4.9 The "is a part of" pattern, Case 2a, non-recursive—(a) generic model and (b) example

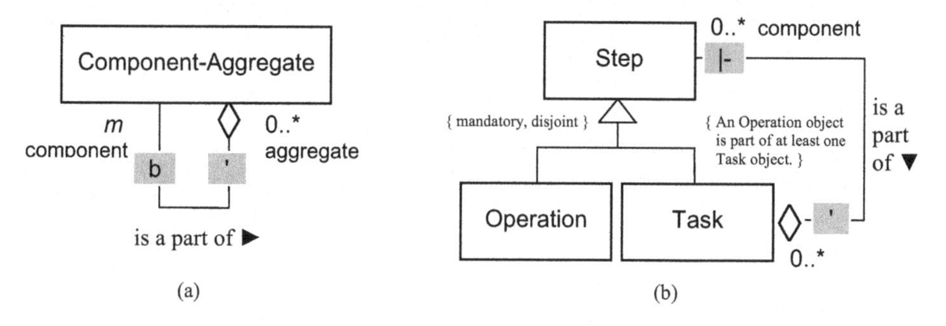

Fig. 4.10 The "is a part of" pattern, Case 2a, recursive—(a) generic model and (b) example

Examples. A Citation object is a part of one or more Document objects but does not exist outside of a Document, <*-to-1..*>? (see Fig. 4.9(b)). A Performance is one of up to twenty performances that can be planned as a part of a Concert, <0..20-to-1..*>?. The same performance can be planned for many concerts.

An Operation is part of one or more Task objects, any of which can be part of one or more other Task objects, etc., until the highest level Task is reached, |-<0..*-

to-0..*>' (see Fig. 4.10(b)). This recursive association is modeled using the Composite design pattern (Gamma et al. 1995), which specializes the Component-Aggregate class into a non-aggregate class, e.g., Operation, and an aggregate class, e.g., Task. A Step must be an Operation or Task but not both. If a Task is deleted (explicitly or implicity), each component is deleted unless it is part of another Task (or cannot be deleted because of an association with some other object, e.g., a Task that was a component is required to define a WorkOrder object). Because of the |- binding, a component can never be explicitly deleted but may be implicitly deleted by disconnecting it from its aggregate.

Variations. Though perhaps rare, an upper bound ≥ 2 can be given for the multiplicity at the aggregate end of the association. This limits the number of aggregates that can contain a component.

Comments. The ?, |-, and ' bindings are necessary for the semantics of this pattern.

The non-recursive version of this pattern is similar to the “is recorded for” Case 2 pattern; however, the “is recorded for” association type lacks the additional semantics of the “is a part of” type, e.g., the notion of whole-part, which cannot actually be expressed in ORN.

4.2.6.3 Case 2b (Shared Aggregation, Dependent Existence, Multiple Associations)

A component object can be part of multiple aggregate objects of the same type or different types, via multiple "is a part of" associations, but can't exist separate from, i.e., independent of, an aggregate object. For each association$_i$ between the Component class and an Aggregate$_i$ class, where $i = 1, 2, \ldots, n$ and $n \geq 2$, the ORN is b<m-to-0..*>' where b includes |- if m_L = 0. Aggregate$_j$ may be the same class as Aggregate$_k$ where $1 \leq j < k \leq n$. The model for the pattern is shown in Fig. 4.11(a).

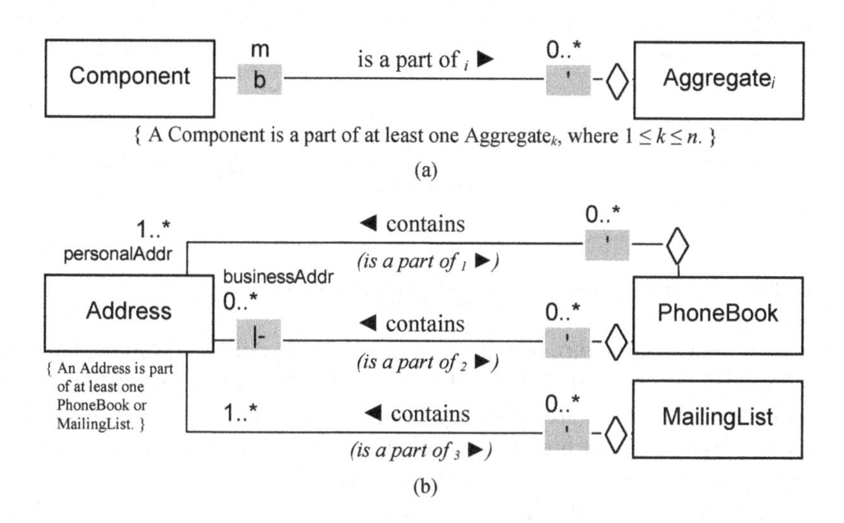

Fig. 4.11 The "is a part of" pattern, Case 2b—(a) generic model and (b) example

Examples. An Address exists only if it is a part of a PhoneBook as a personal address, <1..*-to-*>', or as a business address, |-<0..*-to-*>', or it is a part of a MailingList, <1..*-to-*>' (see Fig. 4.11(b)). A Citation object must be part of one or more Paper or Book objects, |-<*-to-*>'.

Variations. Though perhaps rare, an upper bound can be given for the aggregate end of the association. As an example, assume a participant can be part of many teams as a member but no more than two teams as a leader. Thus a Participant is a part of a Team either playing the role of the leader in one association, |-<0..1-to-0..2>', or that of a member in another association, |-<*-to-*>'.

Comments. The ' binding is necessary for the semantics of this pattern. The |- binding is needed when the lower bound multiplicity for the component class is 0.

The ORN does not prevent independent component objects from being created, but once attached, they become dependent.

A Case 2a pattern can sometimes be used instead of a Case 2b by creating a super class of Aggregate$_i$ classes and associating the Component class to the super class.

This pattern is like the "is recorded for," Case 3 pattern, but again the "is recorded for" association type lacks the whole-part semantics of the "is a part of" type.

4.2.6.4 Case 3a (Composite Aggregation, Independent Existence)

A component object can be part of only one aggregate, i.e., composite, object and can also exist separate from, i.e., independent of, a composite object. If a composite is deleted, all of its components are implicitly deleted. A component can be disconnected from its composite. The ORN for the association pattern is ...-to-0..1>|!, and the model is shown in Fig. 4.12(a).

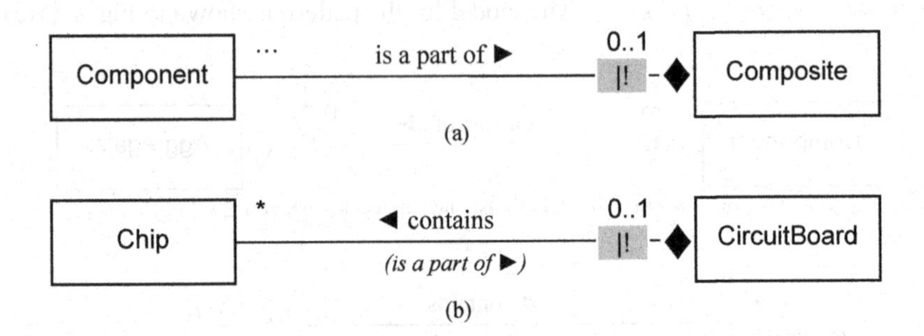

Fig. 4.12 The "is a part of" pattern, Case 3a—(a) generic model and (b) example

Examples. An actual Chip, one that you can hold in your hand, can be part of an actual CircuitBoard, <*-to-0..1>|! (see Fig. 4.12(b)). If a circuit board is deleted, all of the chips that it contains are implicitly deleted; however, a chip can be removed ("de-linked") from a circuit board. Destroying the link does not implicitly delete a Chip object since the binding for the CircuitBoard is |!, versus !.

A Tire is part of an Automobile, <3..4-to-0..1>|!. An original, hardcopy Document is part of a client's File, <*-to-0..1>|!.

Variations. Some revisions to the ORN may be desirable. A ...-to-0..1>|- would allow a composite to be deleted only if all components have been explicitly removed. A StorageBin, for instance, can contain one or more Item objects, |-<0..1-to-*> (inverse <*-to-0..1>|-) and should not be deleted until the items have been removed. A ...-to-0..1>|' would delete a composite if it contains components but would implicitly delete a component only if its existence is not required by some non-"is a part of" association with another object. A ...-to-0..1>' would be like the previous association but would also implicitly delete a component if it is detached from its composite but again, only if its existence is not required by some non-"is a part of" association with another object. If such existence is required, the component would be allowed to be independent. (See the "consists of" association between units in Figs. 1.24 (a) and 2.14 for an example of this latter variation.)

Comments. The |! binding is necessary for the normal semantics of the standard pattern.

4.2.6.5 Case 3b (Composite Aggregation, Dependent Existence)

A component object can be part of only one aggregate, i.e., composite, object and cannot exist separate from, i.e., independent of, a composite object. If a composite is deleted, all of its components, will be implicitly deleted. If a component is detached from its composite, the component is implicitly deleted. A component can be detached from its composite and attached to another via a single operation. Also, a component can be exchanged for another in a composite via a single operation: the original component will be implicitly deleted. The ORN for the association pattern is *b*<...-to-m_L..*1*>! where *b* does not include an X- binding.

If the association is not recursive, m_L = 1 and the model for the pattern is shown in Fig. 4.13(a).

If the association is recursive,. m_L = 0 and the model is shown in Fig. 4.14(a). For a recursive association, the lower bound on the aggregate end is 0 since not all objects can play the role of a component object, i.e., there must be at least one composite that is not a component. (The 0 allows an object to be created that is not part of another object. Such an object must be a composite.)

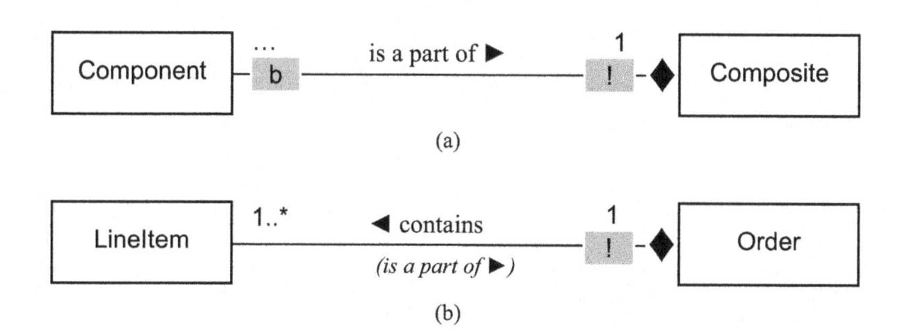

Fig. 4.13 The "is a part of" pattern, Case 3b, non-recursive—(a) generic model and (b) example

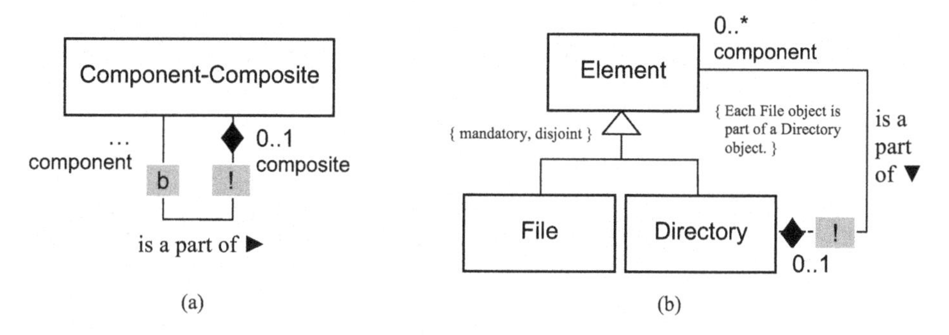

Fig. 4.14 The "is a part of" pattern, Case 3b, recursive—(a) generic model and (b) example

Examples. A LineItem is a part of an Order, <1..*-to-1>! (see Fig. 4.13(b)). A Choice is a part of a Menu, ?<2..*-to-1>! (which is defined by at least two choices).

In a file system, a File is a part of a Directory, which itself can be part of a Directory, etc., until a highest level or root Directory is reached, <*-to-0..1>! (see Fig. 4.14(b)). This recursive association is modeled using the Composite design pattern (Gamma et al. 1995). A file system Element must be a File or Directory but not both. A component of a Directory can be moved from one directory to another, but cannot exist on its own.

Variations. Some revisions to the ORN may often be desirable. A …X-< …-to-m_L..1>! only allows a component to be detached from its composite (and implicitly deleted) by exchanging it with a new component in a single operation. A <…-to m_L..-1>X-|! only allows a component to be detached from its composite by reattaching it to another composite in a single operation. Perhaps more commonly desired, a …X-< …-to- m_L..1>X-|! makes an attachment permanent. A Room is a permanent part of a Building, X-<*-to-1>X-|!.

Comments. The ! binding is necessary for the recursive version of this pattern.

4.2.7 "is a" pattern

When a programming language or a DBMS does not support an "is a" type of relationship, then the "is a" must be implemented as an association. In this situation, a *subclass object* (though a separate object) "is a" *superclass object.* When a subclass object is deleted, the related superclass object must be implicitly deleted; likewise, when a superclass object is deleted, any related subclass object must be implicitly deleted. Thus, the related objects are essentially treated as the same object.

4.2.7.1 Case 1 (Optional Subclasses)

Membership of a superclass object in one or more of its subclasses is optional. Destroying the link between a subclass object and its related superclass object removes the superclass object from the subclass by deleting the subclass object. For each association $_i$ between a SubClass $_i$ and its SuperClass, where $i = 1, 2, \ldots, n$ and $n \geq 1$, the ORN is |!<0..1-to-1>!. The model for the pattern is shown in Fig. 4.15(a).

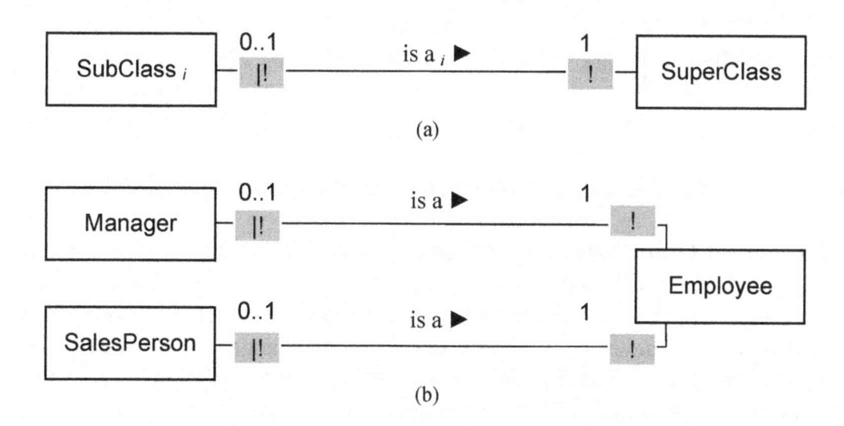

Fig. 4.15 The "is a" pattern, Case 1—(a) generic model and (b) example

Examples. A Manager is an Employee, |!<0..1-to-1>!, and a SalesPerson is an Employee, |!<0..1-to-1>! (see Fig. 4.15(b)). Not all employees are managers or salespersons. Because a |! binding is given for Manager instead of a ! (or |!X!) binding, if the link between a Manager object and an Employee object is explicitly destroyed, the related Employee object is not implicitly deleted (since explicit default destructibility is applicable), but the Manager object is implicitly deleted because of the X! binding, implied in the ! binding, for Employee. In effect, the employee is no longer a manager.

A LabClass and GenEdClass have |!<0..1-to-1>! "is a" associations with the class Class. A SpecialOrder "is a" Order, |!<0..1-to-1>!.

Comments. The |! binding is necessary for the semantics of this pattern.

4.2.7.2 Case 2 (Mandatory Subclasses)

Membership of a superclass object in one or more of its subclasses is mandatory. Once a link is established between the a subclass object and superclass object, it cannot be explicitly destroyed. For each association$_i$ between a SubClass $_i$ and its SuperClass, where $i = 1, 2, \ldots, n$ and $n \geq 2$, the ORN is |!X-<0..1-to-1>X-|!. The model for the pattern is shown in Fig. 4.16(a).

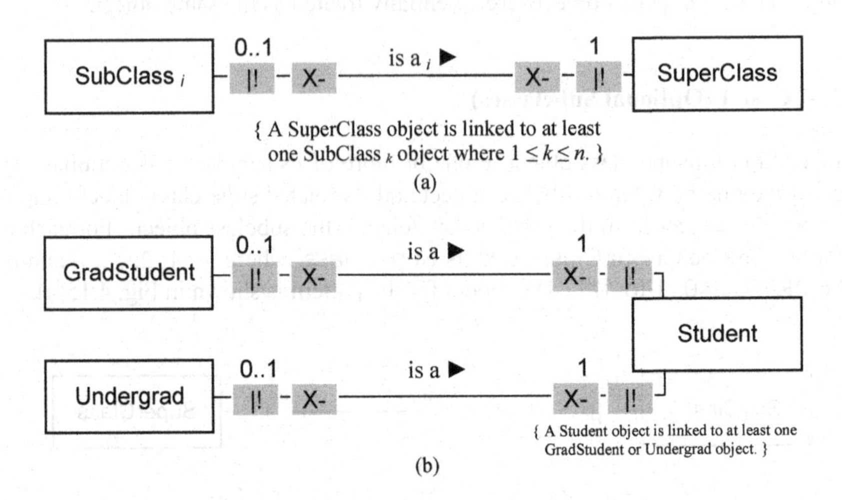

Fig. 4.16 The "is a" pattern, Case 2—(a) generic model and (b) example

Examples. A GradStudent is a Student, and an UnderGrad is also a Student, both |!X-<0..1-to-1>X-|!, and a student must be either a graduate student or undergrad (see Fig. 4.16 (b)). A link between an UnderGrad object, for instance, and a Student object can never be explicitly destroyed or changed once created. (If an UnderGrad becomes a GradStudent, a new Student object must be created.)

A Button is a ScreenControl (as are many other types of screen controls), |!X-<0..1-to-1>X-|!. A screen control object can only be instantiated as one of its subtypes and once instantiated remains as always this subtype.

Comments. The |! and X- bindings are necessary for the semantics of this pattern.

4.3 Patterns in Database Modeling and Implementation

This section presents two examples to demonstrate the use of multiple association patterns in database models and their implementation. The first example shows how the "is associated by" pattern is used to remodel many-to-many associations, which may have been modeled using other association patterns. Such remodeling is necessary to implement many-to-many associations in relational databases and many-to-many associations having association attributes, no matter the implementing system. The remodeling is also necessary, and can be easily extrapolated, to implement n-ary associations where $n > 2$. The second example shows how multiple association patterns can be used to model part of a company database. This model is implemented in two ways, first mapping it to a DDL for an object DBMS and second to a DDL for a relational DBMS. Both DDLs support ORN, meaning that all of the semantics inherent in the association patterns are automatically enforced by the DBMS.

Fig. 4.17 shows how a many-to-many association between songs and CDs with attribute trackNo must be remodeled into two one-to-many associations using the "is associated by" pattern. The many-to-many association fits both the "is a part of" (Case 1) and "is defined by" patterns. The generic association names for all applicable patterns are given where appropriate in the class diagrams. The ovals and arrows indicate how the multiplicities and bindings for the two classes in the original association map to those for the new associating class, here Track. If deletion of an associating class object is to be equivalent to explicit destruction of a link in the original association, then in this association the explicit binding and implicit binding on each related class must be the same.

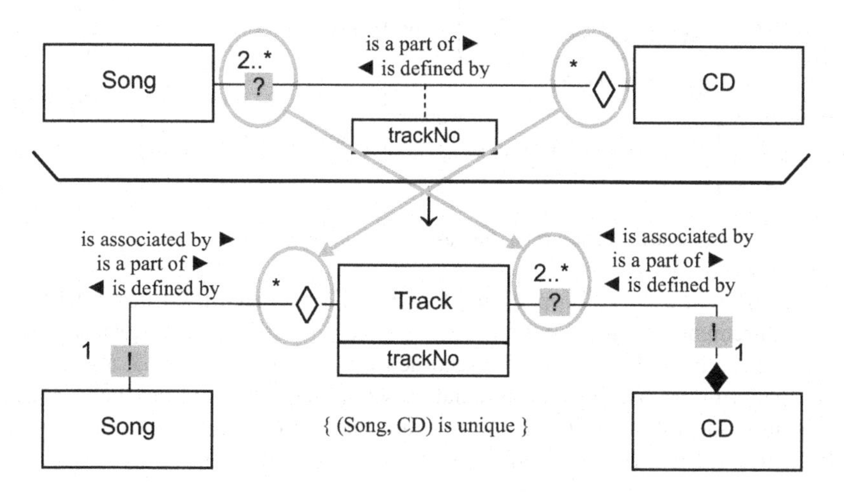

Fig. 4.17 Converting a many-to-many association using the "is associated by" pattern.

The semantics of the many-to-many association are preserved in the remodeling. For example, if a song is deleted, all of the related tracks will be implicitly deleted, and if any of these tracks represent one of only two tracks on a CD, the CD is implicitly deleted. Note that half of the "is a part of" (Case 1—Shared Aggregation, Independent Existence) association between Song and CD has been automatically and properly upgraded to an "is a part of" (Case 3b—Composite Aggregation, Dependent Existence) between Track and CD (since a |! is implied by the !).

Fig. 4.18 models some classes and associations that may be applicable to the database requirements of a company. The company is hierarchically organized into units. An employee works for a unit—e.g., corporate office, product support, or

DBMS support—and has a job assignment that describes, in more detail than that of the employee's job description, the expectations for the job currently assigned to the employee, i.e., its duties and responsibilities. Multiple employees can share the same job assignment, and employees are often given new assignments. Occasionally, the company hires temporary workers. Job assignments may also be used to describe the jobs for which these workers were specifically hired. Job assignments no longer applicable to any employee or temporary should not be recorded. The company encourages its employees to car pool and keeps track of existing carpools by having employees register them and sign up for them. Occasionally, to keep the records up-to-date, it surveys employees on their carpooling. Carpools should be removed when ridership falls below two.

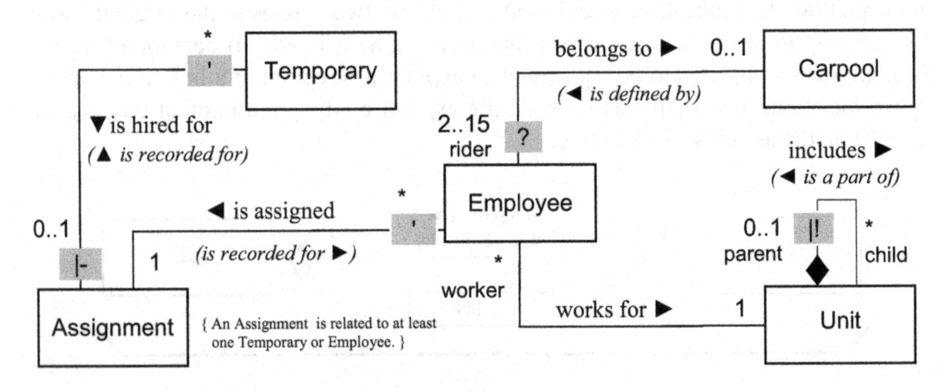

Fig. 4.18 Partial company database using four association patterns

Three different association patterns, four in all, have been used in modeling the above requirements as shown in Fig. 4.18. They are identified by the generic pattern names given in parentheses. The "is recorded for" pattern is a Case 3, and the "is a part of" pattern is a Case 3a. The association semantics resulting from the use of patterns include the following (with the relevant generic object names, bindings, and multiplicities given in parentheses):

- If an employee (defining object) is deleted or removed from a carpool (defined object) and if the employee is one of just two riders in the carpool, the carpool will be implicitly deleted (? and 2).
- If an employee (primary object) or temporary (primary object) is deleted or if a job assignment (supporting object) is taken away from an employee or temporary, then an attempt is made to implicitly delete the related job assignment (' for Employee and Temporary). This deletion is successful only if the assignment is not that of another employee (implicit default binding and 1 for Assignment in "is assigned") or temporary (|- for Assignment in "is hired for").
- If a unit (composite) is deleted, all descendant (component) units are implicitly deleted (|!); however, the deletion of the primary unit will fail if any units have employees working for them (default implicit destructibility and 1).

The semantics of associations modeled based on association patterns can be easily implemented when a DBMS supports ORN. All of the association examples given in this chapter were implemented in such an object DBMS, specifically OR+, and they were tested to verify that their ORN semantics were correctly maintained by the DBMS. In fact, the reader can replicate these tests using the ORN Simulator, which was discussed in Chapter 3. In addition to the pattern matching associations given in this paper, many other such associations were successfully implemented and tested in a relational DBMS using ORN Additive, which is discussed in Chapter 6. Fig. 4.19 shows how the associations modeled in Fig. 4.18 were implemented in an object database using OR+, which is discussed in Chapter 7. Fig. 4.20 shows how they were implemented in a relational database using ORN Additive.

```
...
class Unit {
        d_String              name;
        relationship Unit     parent inverse children <*-to-0..1>|!;
                              // the "relationship" keyword is optional
        Set<Unit>             children inverse parent;
        Set<Employee>         workers inverse unit <1-to-*>;
        ...
};
class Carpool {
        d_String              id;
        Set<Employee>         riders inverse carpool;
        ...
};
class Assignment {
        d_String              code;
        Set<Temporary>        temporaries inverse assignment;
        Set<Employee>         employees inverse assignment;
        ...
};
class Employee {
        d_String              ssn;
        Carpool               carpool inverse riders ?<2..15 -to-0..1>;
        Assignment            assignment inverse employees '<*-to-1>;
        Unit                  unit inverse workers;
        ...
};
class Temporary {
        d_String              ssn;
        Assignment            assignment inverse temporaries '<*-to-0..1>|-;
        ...
};
...
```

Fig. 4.19 ODDL implementation of model given in Fig. 4.18

In Fig. 4.19, the ORN for each modeled association is given for one of the two *relationship traversal paths* (Cattel et al. 2000), or *object-based attributes*, that implements it. Here, the DDL is the ODDL of OR+, which is compatible to the Object Definition Language (ODL) of the ODMG standard (Cattel et al. 1997). The ODML of OR+ provides an APL for inserting, updating, and deleting objects in the ODDL-defined object database. As these operations are performed, the semantics of the ORN, and thus those of the association patterns, are automatically maintained by the object DBMS. Database developers are spared the effort of having to develop and maintain the complex methods that would otherwise be required to implement these semantics.

```
...
CREATE TABLE Unit (
  name          VARCHAR(20)  PRIMARY KEY
...
);
CREATE TABLE Carpool (
  id            VARCHAR(8)  PRIMARY KEY
...
);
CREATE TABLE Assignment (
  code          CHAR(3)  PRIMARY KEY,
...
);
CREATE TABLE Employee (
  ssn           CHAR(11)  PRIMARY KEY,
  ...
  carpoolId     VARCHAR(8)  CONSTRAINT belongsTo REFERENCES Carpool(id),
--+<> belongsTo ?<2..15-TO-0..1>;
  assignCode    CHAR(3)  CONSTRAINT isAssigned REFERENCES Assignment(code),
--+<> isAssigned '<*-TO-1> ON UPDATE CASCADE;
  unitName      VARCHAR(20)  CONSTRAINT worksFor REFERENCES Unit(name)
--+<> worksFor <*-TO-1>;
);
CREATE TABLE Temporary (
  ssn           CHAR(11)  PRIMARY KEY,
  ...
  assignCode    CHAR(3)  CONSTRAINT isHiredFor  REFERENCES Assignment(code)
--+<> isHiredFor '<*-TO-0..1>|- ON UPDATE CASCADE;
);

ALTER TABLE Unit ADD
  parent        VARCHAR(20)  CONSTRAINT includes REFERENCES Unit(name);
--+<> includes <*-to-0..1>|!;
...
```

Fig. 4.20 T-SQL and ORN Additive implementation of model given in Fig. 4.18

In Fig. 4.20, the ORN for each modeled association is given for the foreign key constraint that implements it. (In a relational context, classes essentially become tables and objects become rows.) Here, the DDL is T-SQL of SQL Server (Microsoft Inc. 2005) supplemented by the --+<> statements of ORN Additive. After the T-SQL is executed to create the database tables and constraints, the --+<> statements are processed by an ORN Additive utility. This utility automatically generates the T-SQL code that when executed creates the meta-data, triggers, and stored procedures required to maintain the semantics of the ORN, and thus those of the association patterns. The triggers are invoked when T-SQL queries insert, update, and delete rows in the *ORN-related tables*. Here again, database developers are spared the effort of having to develop and maintain the complex code needed to implement the association patterns. In all, 3,976 lines of T-SQL code were generated by ORN Additive in order to implement the associations given in Fig. 4.20 (though granted, the code is well-commented and some code is generic rather than totally customized).

4.4 Associations That Don't Conform to a Pattern

Before concluding this chapter, we should examine some associations that do not conform to any of the association patterns described in Section 4.2. Obviously, many such associations exist.

The first example is the "works for" association between the classes Employee and Unit in Fig. 4.18. This association fits none of the patterns defined in Section 4.2. As is typical of many such associations, it involves a relationship between classes that are relatively independent and on "equal footing," which usually means that they are both strong entity types, as defined by Chen (1976).

Another example is shown in Fig. 4.21. A professor "advises" zero or more students. A student may or may not have been assigned a professor as an advisor. This association, like the "works for" association involves strong entity types. Also, like the "works for" association, it is defined with only default bindings, which is also common for associations that fit none of the patterns defined in Section 4.2.

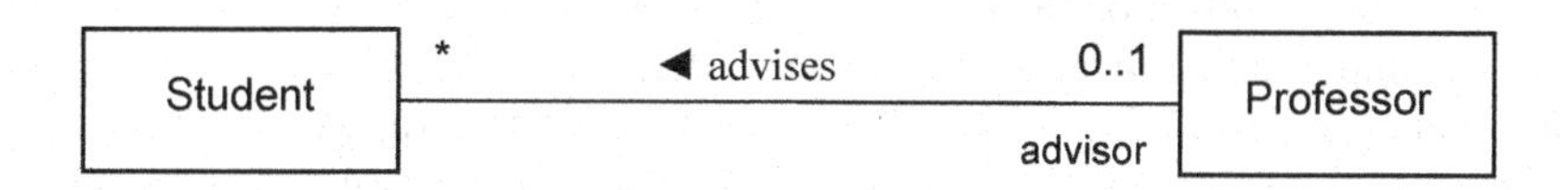

Fig. 4.21 Example of a non-pattern conforming association

Fig. 4.22, on the other hand, shows an example of a non-pattern conforming association that is defined with a non-default binding. An object class defined in a CASE tool may depend on one or more other object classes. The |- binding means that a class cannot be deleted if any other classes are dependent on it. All dependent classes must first be deleted or their dependencies must be removed.

But is this really a non-pattern conforming association? Or, does it indicate the existence of another potential association pattern, a generic "is dependent on" association type between objects?

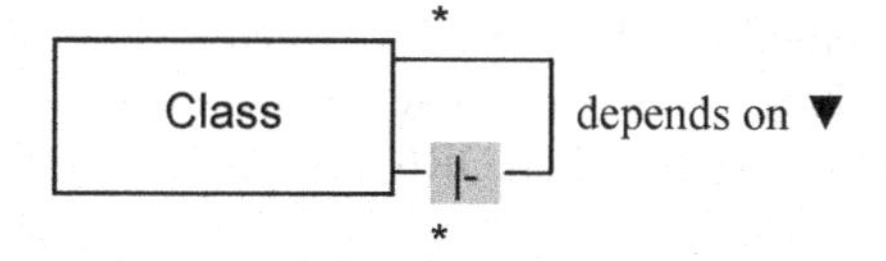

Fig. 4.22 Example of a non-pattern conforming association with a non-default binding

4.5 Conclusion

Students often struggle in Database Management classes to recognize and properly model the associations that are present in real or sample database applications. What may seem commonplace and obvious to the experienced instructor is anything but to a student. These observations were a major motivation for developing association patterns. These patterns allow the experience of the more expert data modeler to be more readily transferred to the student or novice practitioner, thus making data modeling more of a science and less of an art. In this regard, they are like analysis patterns; however, unlike these patterns, they are application-independent and thus don't require the more extensive study of the common requirements of numerous business-related applications in order to gain modeling expertise.

Another related motivation for developing association patterns was to identify and more precisely define the semantics, especially the multiplicities and referential actions (or bindings), inherent in the common types of associations that appear in many applications. In this regard, association patterns distinguish, for example, the semantics of the "is defined by" association type from those of the "is associated by" type, allowing such types to be more easily recognized in applications and the proper multiplicities and bindings to be more readily assigned. Association patterns also define the precise semantics of the common variations that occur for some association types. For example, the "is a part of" association type was dissected into five association patterns with variations, all of which more precisely define the semantics of the shared aggregation and composition associations in UML.

Though not a motivation, a significant outcome from developing association patterns is an important application and, to some degree, a validation of the work done by this author in developing ORN. The association patterns validate the bindings of ORN by showing that all of them are necessary. The bindings are also sufficient to a large extent in providing the referential actions needed for these association patterns. Admittedly, they may not be as sufficient for other association patterns that may be described in the future. Also, a certain lack of sufficiency is evident from the added constraints given within braces for some of the current patterns (see for instance Fig. 4.16). It is extremely doubtful, however, that any declarative scheme for describing relationship semantics will ever be totally sufficient.

Nevertheless, the use of ORN in developing association patterns makes a stronger case for including more declarative association semantics, like ORN, into UML and into the DBMS, as is argued throughout this book. Also, experience in modeling and implementing numerous classes and associations has shown that the productivity of developing and maintaining database systems is increased significantly when association patterns are used to model databases and when these patterns are easily mapped to and implemented in a DBMS—i.e., when a more model-driven approach is used (Mellor et al. 2003). This is possible when an ORN-extended class diagram is used for database modeling and when ORN is supported by the DBMS.

Chapter 5

Comparing ORN to Similar Declarative Schemes

In this chapter we compare ORN to other declarative schemes for describing association semantics. We examine schemes that have been proposed by others and ones that have been generally adopted in current DBMSs. We also examine property schemes that have been proposed for whole-part, i.e., "is a part of" type, relationships. Our purpose is to better assess the relative simplicity, power, consistency, and uniqueness of ORN and its scope. The chapter updates and extends the comparisons made in Ehlmann and Riccardi (1996).[1]

In making the comparisons, we should recognize the following:

- A declarative scheme can never provide for all of the semantics that may be relevant to a relationship. A language for specifying integrity rules with general conditions and triggers is more powerful in expressing relationship semantics but is also more verbose and less efficient in dealing with the specialized integrity rules that define common types of relationships (Date 1981, Stonebraker et al. 1990). Thus any declarative scheme represents a compromise and must be judged on its capacity to provide the strongest semantics for the most common relationship types with a minimum of syntax and conceptual complexity.
- Most semantic modeling schemes, including ORN, are developed mostly on pragmatic grounds (Rundensteiner et al. 1994). Although syntax and concepts can be invented to cover some apparent generic cases, no rigorous proof of "completeness" can be given. Rather, the utility of a new scheme is shown by giving many concrete examples of cases, e.g., real relationships, that are handled adequately by the scheme, emphasizing those not adequately handled by existing schemes.
- ORN is defined at a conceptual level in terms of an ER or class diagram. Some of the schemes to which ORN is compared are defined only in terms of a specific logical model, the relational model or a type of object model. Although ORN can be adapted to both relational and object models, as is shown in later chapters, equivalent ORN specifications, when given, are given at a conceptual level.

The remainder of this chapter is organized into four sections. Section 5.1 compares ORN to related declarative schemes for the relational data model and for relations in a proposed object model. Section 5.2 compares ORN to the declarative capabilities for describing relationship semantics that generally exist in an ODBMS. Section 5.3 compares ORN to related declarative schemes that have been proposed for the ER model. Section 5.4 describes how ORN relates to efforts to better define

[1] Portions re-used, with permission, from "A comparison of ORN to other declarative schemes for specifying relationship semantics," *Information and Software Technology* 38, 7. © 1996 Elsevier.

B.K. Ehlmann, *Object Relationship Notation (ORN) for Database Applications*,
Advances in Database Systems 39, DOI 10.1007/978-0-387-09554-7_5,

the semantics of whole-part relationships via numerous properties. It also compares ORN to a proposed multi-dimensional framework and notation for differentiating aggregation, composition, and normal association semantics in UML class diagrams. Section 5.5 concludes with some summary assessment remarks regarding ORN.

5.1 Schemes for Relational Databases and Object Relations

ORN is comparable to schemes for defining relationship semantics in the relational model and in proposed object models that include relations. For the relational model, Date (1981) and Markowitz (1990) propose declarative referential integrity rules and assume an optional null constraint rule for foreign keys. The rules proposed by Date (1981) and the null constraint are part of the SQL standard (ANSI 2008). Albano et al. (1991), like Rumbaugh (1987, 1988), includes a relation structure in an object model to represent an association and proposes referential, surjectivity, dependency, cardinality, and constancy constraint rules to describe associations. This section first shows that ORN provides semantic equivalencies to all of the above mentioned rules, though the packaging is radically different. It then discusses some additional relationship semantics that ORN provides and some that it does not.

Before proceeding, mention should be made that Rundensteiner et al. (1994) proposes still another declarative scheme for relationship semantics. While this scheme includes an extensive array of set restrictions, it was not included in the comparisons here since it does not provide any rules for what the database system should do when restrictions are violated, an important aspect in defining relationship behavior.

5.1.1 Null constraint

A *null integrity constraint* on an attribute in a relational database states that its value cannot be null. The availability of this constraint on foreign keys is assumed in Date (1981) and Markowitz (1990). Disallowing null values for a foreign key is equivalent to an ORN *<multiplicities>* specification of …-to-1 between the object type, or class, represented by the referencing relation and that represented by the referenced relation. The "…" implies that any *<multiplicity>* may be given. Allowing null values for a foreign key is equivalent to declaring a …-to-0..1 association.

Unit(unitNo, name, ...);
Employee(ssn, name, ... , unitNo NOT NULL), FOREIGN KEY unitNo REFERENCES Unit;

Fig. 5.1 Relations Employee and Unit implementing the "works for" association

For example, Fig. 5.1 shows relations implementing the "works for" relationship between employees and units. A null constraint on the unit number attribute, unitNo, in the relation Employee in Fig. 5.1 is equivalent to a ...-to-1 association between employees and units.

5.1.2 Key constraints

Relationship multiplicities are also indirectly specified for relations by the presence or absence of key integrity constraints on foreign keys. These *key constraints*, realized via the declaration of candidate and primary keys, which must be unique, are integral to the relational model. In Albano et al. (1991) they provide cardinality constraints for object class relationships represented as relations.

Declaring a foreign key as a candidate key in a relation is equivalent to declaring a 0..1-to-... association between the object class represented by the referencing relation and that represented by the referenced relation. Not declaring a foreign key in a relation as a candidate key is equivalent to declaring a *-to-... association between these relations. A null constraint is assumed for a candidate key that is made a primary key. The unitNo attribute in relation Employee in Fig. 5.1 is not declared as a candidate key and has a null constraint as shown. This is equivalent to a *-to-1 association between employees and units. If the unitNo attribute in relation Employee was declared as a candidate key, then no two employees could have the same unitNo value, i.e., work for the same unit, and thus the association between employees and units would be 0..1-to-1. Finally, assume the following two relations:

W-4(ssn, noExemptions, ...), FOREIGN KEY ssn REFERENCES Employee;
Employee(ssn, name, ...), FOREIGN KEY ssn REFERENCES W-4, ...;

Here, the association between employees and W-4s would be 1-to-1.

In Albano et al. (1991) relationship constraints are only defined for *object attributes* (i.e., non-null, single-valued, object-based attributes) that are *components* (i.e., foreign keys) of an *association*, where the association represents an *n*-ary relationship among component, referenced classes and is similar to a relationship relation in the ER model. Declaring these components as elements of a key implies a many-to-many relationship between all component classes. Not including a component as an element of a key implies a one-to-many relationship between the non-key component class and the key component class(es). Declaring two components as separate keys would indicate a one-to-one relationship between these component classes.

In ORN a unique constraint given for an associating class—a class that defines an association as an object, like the Assignment class in Fig. 5.2—provides the same semantics. With ORN, however, most binary associations are not represented as associating classes and multiplicities are declared directly (not indirectly via unique key constraints). Nevertheless, in Fig. 5.2 the one-to-many association between units and employees can be specified by excluding the Unit object from the unique key for Assignment and placing the constraint "{ related Employee object is unique }" near the Assignment class. The multiplicity for this class is then 0..1 in its association with the Employee class and * in its association with the Unit class, thus specifying a 0..1-to-* association between units and employees.

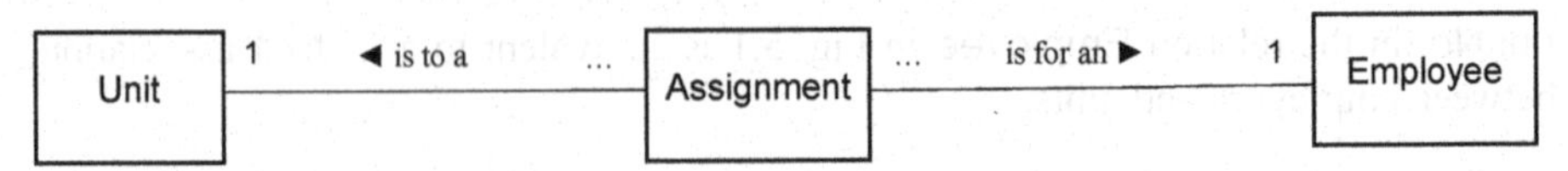

Fig. 5.2 Class diagram with "works for" relationship represented using a separate associating class

5.1.3 Referential integrity rules

Referential integrity in relational databases means that every non-null foreign key value must match a primary key value in the referenced relation. In an object database, referential integrity means that the values of all non-null, object-based attributes must reference objects in the database—that is, no dangling object references can exist. Date (1981) proposes *referential integrity rules* that specify what should be done when an operation is attempted that would violate referential integrity in a relational database. These rules are given for each foreign key and provide additional relationship semantics. In Markowitz (1990) the rules are scaled back for relational databases generated from an ER model, and in Albano et al. (1991) they are reduced and repackaged for relations representing relationships in the proposed object data model. The following paragraphs discuss ORN equivalencies to Date's and Albano's referential rules, which are similar to those in standard SQL (ANSI 2008).

The *inserting restricted rule* is automatic and states that a tuple with a non-null, unmatched foreign key value cannot be inserted into a relation. For example, a tuple with a unitNo value of 03 cannot be inserted into an Employee relation, as defined in Fig. 5.1, if there is no tuple with unitNo 03 in the Unit relation. The object database equivalent of the inserting restricted rule is assumed by the in and owned_by referential constraints given in Albano et al. (1991). (The are referential constraint in Albano et al. (1991) assumes an "inserting cascades" rule but seems relevant only to the implementation of an "is a" relationship.) These inserting rules are only needed because of the way relationships are represented in relational and object databases, i.e., by references. They are not relevant to ORN, which describes relationships at a conceptual level.

The UPDATING rules specify what should be done relative to a particular foreign key when a primary key value is updated in a tuple of the referenced relation. UPDATING rules are not relevant to object databases since objects are not referenced by primary key values. Instead, they are referenced by object ids, which cannot be updated. UPDATING rules are also not relevant to ORN since they do not specify relationship semantics but instead pertain to maintaining the integrity of the relationship representation in a relational database.

The DELETING rules (i.e., the ON DELETE referential actions in SQL) specify what should be done when a tuple is deleted from a referenced relation.

The RESTRICTED option (i.e., RESTRICT in SQL) means that the delete fails if matching tuples exist in the referencing relation. For example, in Fig. 5.1 deletion of a Unit tuple with unitNo 03 fails if an Employee tuple exists with unitNo 03.

RESTRICTED is the default DELETING rule in Date (1981) but not in standard SQL. It corresponds to the in referential constraint in Albano et al. (1991).

In ORN a |- binding on the class represented by referenced relation provides the equivalent constraint. For example, a <*-to-0..1>|- association between employees and units means that a unit cannot be deleted if it has an employee.

The NULLIFIES option (i.e., SET NULL in SQL) means that foreign keys in matching tuples of the referencing relation are set to null when a tuple is deleted from a referenced relation. For example, deletion of a Unit tuple with unitNo 03 sets the unitNo to null in all Employee tuples having unitNo 03. (This option would obviously conflict with a null constraint on the foreign key.) DELETING NULLIFIES has no corresponding referential constraint in Albano et al. (1991) since object attributes in relations cannot be null.

In ORN the equivalent constraint is provided by a default implicit destructibility binding and 0..1 multiplicity for the class represented by the referenced relation. For example, if the relationship between employees and units is <*-to-0..1>, then deletion of a unit causes the implicit destruction of all association links involving employees.

The CASCADES option (i.e., CASCADE in SQL) means that matching tuples in the referencing relation are deleted when a tuple is deleted from a referenced relation. For example, deletion of a unit tuple with unitNo 03 deletes all Employee tuples having unitNo 03. DELETING CASCADES corresponds to the owned_by referential constraint in Albano et al. (1991).

In ORN, the equivalent constraint is provided by a |! binding for the class represented by the referenced relation. For example, if the relationship between employees and units is <*-to-0..1>|!, then deletion of a unit causes all related employees to be deleted.

A NO ACTION option, which is not provided by Date (1981) but is by SQL, means that no action is taken by the DBMS if matching tuples exist in the referencing relation. On transaction commit, however, if such tuples still exist, an exception results. So, this option assumes that the application will somehow fix the problem prior to commit. For example, deletion of a Unit tuple with unitNo 03 results in no action if an Employee tuple exists with unitNo 03; however, if a referential integrity violation is detected on the unitNo foreign key at transaction commit, an exception results. NO ACTION is the default ON DELETE action in standard SQL.

In ORN, an implicit default destructibility binding and 1 multiplicity on the class represented by the referenced relation provides a similar constraint in that the check for multiplicity violation is deferred until the commit of the complex object operation. For example, a <*-to-1> association between employees and units implicitly destroys a link to a related employee on deletion of a unit; however, the deletion will fail on commit of the complex object operation if the 1 multiplicity is still violated. Setting RXC_mode before deleting a referenced object may also provide the system action desired in that checks for lower bound violations are deferred until the commit of the application-defined transaction.

5.1.4 Surjectivity constraints

Referential integrity and its associated rules do not address a very common relationship constraint. In relational terms, this constraint for a foreign key asserts that a tuple in the referenced relation must have at least one matching tuple in the referencing relation. For example, a Unit tuple must have at least one Employee tuple with matching unitNo, which implements a policy that every unit have at least one employee. The constraint may be unusual for the employees and units relationship, but for other relationships it is not, e.g., a track in a detector is defined in terms of, i.e., must have, at least one subtrack. As with referential integrity, this constraint must have associated rules specifying what must be done when an operation is attempted that would violate it. In Albano et al. (1991) the onto and owns clauses provide these rules and are collectively called *surjectivity constraints*—a term derived from viewing relationships as functions (see Section 9.4).

Before discussing ORN equivalencies to onto and owns, we need to clarify some things about surjectivity and our discussion. In Albano et al. (1991), surjectivity is defined only for components of associations. An association corresponds to a relationship relation in the ER model or a class that represents an association in a class diagram. For purposes of comparison, the discussion below treats surjectivity only in the context of such an associating class. Examples are based on the units and employees relationship, as represented in Fig. 5.2 by the class Assignment. As indicated in the previous paragraph, however, surjectivity is relevant for relationships represented as foreign keys in relational databases, e.g., unitNo in relation Employee, and as object-based attributes in object databases, e.g., a Unit object-based attribute in class Employee. ORN provides surjectivity for such attributes in general, not just for those in associating classes, like Assignment.

As with referential integrity, an inserting restricted rule is automatic for surjectivity and is assumed when onto or owns is specified for an association component. For example, if such is the case for the Unit component of an Assignment association, then a Unit object x cannot be inserted into the Unit component class unless a tuple referring to x exists in the Assignment association. The equivalent of surjectivity in ORN is specified by a lower bound multiplicity of 1 for the associating class. For example, if the relationship between units and assignments is 1-to-1..*, then a Unit object cannot be inserted into a database unless it is linked to an Assignment object.

In Albano et al. (1991), whether the onto or the owns clause is specified for a component of an association indicates what should be done when a tuple is deleted from an association.

Specifying onto indicates that the delete fails if the tuple deleted is the only one referring to an object of the component class (unless this object is also removed in the same transaction). For example, an onto specified for the Unit class component of an Assignment association would mean that deletion of an Assignment tuple referencing a Unit object x would fail if it is the only Assignment tuple referencing x (unless x is also removed). In ORN the default implicit destructibility binding for the associating class provides an equivalent constraint. For example, if the relationship between units and assignments was <1-to-1..*>, then an Assignment object whose related Unit is *x* cannot be deleted if it is the only Assignment object related to *x*, i.e., if its deletion (actually, the required implicit relationship destruction) violates the 1..* multiplicity.

Specifying owns indicates that the delete of an association tuple causes the automatic deletion of the component class object if the tuple deleted is the only one referring to the object. For example, an owns specified for the Unit class component of the Assignment association would mean that deletion of an Assignment tuple referencing Unit object *x* causes the deletion of *x* if it is the only Assignment tuple referencing *x*. In ORN the |? binding for the associating class provides an equivalent constraint. For example, if the relationship between units and assignments was <1-to-1..*>|?, then the deletion of the only Assignment object related to an Employee object *x* causes the deletion of *x* since the deletion violates the 1..* multiplicity.

5.1.5 Additional relationship semantics

As previously stated, the default implicit destructibility binding along with a lower bound multiplicity of 1 for an associating class provides an equivalent in ORN to the onto surjectivity constraint. The |- binding provides an additional and stronger surjectivity rule for deletion. A <1-to-1..* >|- instead of a <1-to-1..*> association between units and assignments, for example, means that an assignment can *never* be deleted. (A !<1-to-1..* >|- association would delete an assignment implicitly as a result of deleting the related unit or destroying the association between a unit and an assignment.)

The constancy constraint proposed in Albano et al. (1991) makes a relationship indestructible. In ORN, if the associations between units and assignments was <1-to-* >- (i.e., <1-to-*>|-X-) and that between employees and assignments was <1-to-0..1>-, then an Assignment object can never be deleted and the relationship between a unit and an employee via an Assignment object once established can never be destroyed or changed.

In OR+ (see Chapter 7), an X- constraint is implemented for a class independent of any association. (This X- is not the X- binding.) It indicates that class objects once created cannot be explicitly destroyed, i.e., deleted. If X- was specified for the Assignment class and if the associations between units and assignments was !<1-to-* >X-) and that between employees and assignments was !<1-to-0..1>X-, then again the relationship between an employee and a unit via an assignment object once established can never be explicitly destroyed; however, it is now implicitly destroyed when the related employee or project is deleted.

Suppose an employee was required to work for more than one unit. To make this a bit more realistic, suppose an employee was required to be involved in at least three projects. This many-to-many relationship is modeled in Fig. 5.3. The relationship between assignments and employees is <3..-to-1>!. Here, the deletion of one of only three assignments for an employee would fail. If the association was |?<3..-to-

1>!, then the deletion of one of only three assignments for an employee would delete the employee. These types of surjectivity constraints are possible only when a declarative scheme for specifying relationship semantics is based on multiplicities. The "must have at least one" constraint for surjectivity as defined in Albano et al. (1991) should be seen as just a special case of a more general surjectivity.

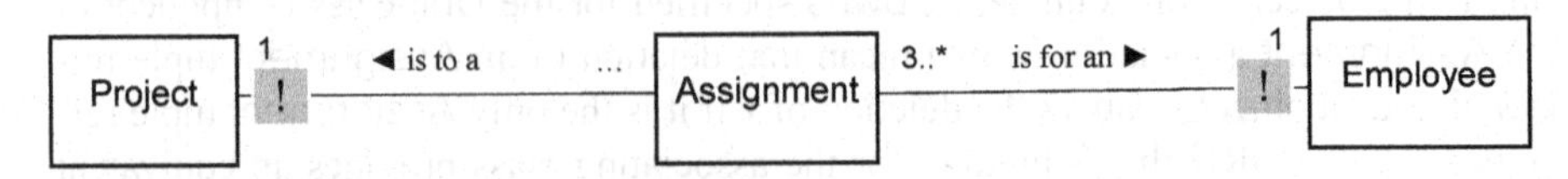

Fig. 5.3 Class diagram for many-to-many relationship between projects and programmers

In Albano et al. (1991), different surjectivity constraints can be given for the subclasses of a component class in an association. For example, the Employee component in the Assignment association could have no surjectivity constraint while a Programmer, who "is a" Employee, could have an onto constraint. This would implement a policy that allows an employee in general to have no assigned project but requires a programmer to have at least one. This capability can be achieved using ORN by defining an association specifically for a subclass that has the desired multiplicity and binding on the associationg class, e.g., a <1..*-to-1> association between Assignment and Programmer.

Finally, the semantics of the ' binding in ORN, although seemingly similar to the ! binding, are quite unique. These semantics, in combination with appropriate multiplicities and other bindings, facilitate shared "subordinate" objects among "prime" objects from either the same or different classes and implement an automatic garbage collection mechanism.

Fig. 5.4 Class diagram for association between employees and doctors

To illustrate, consider the relationship between employees and doctors in Fig. 5.4. An employee optionally "records" a doctor as his or her primary physician. Recorded attributes for a doctor may be name, office address, and phone number. The relationship is '<*-to-0..1>|-. The |' aspect of the ' binding means that if an employee *x* is deleted, any association with a doctor is implicitly destroyed and an attempt is made to implicitly delete the related doctor. If the doctor is also recorded for another employee, i.e., if there exists an association between the doctor and another employee, the doctor is not deleted since the |- binding does not allow implicit destruction of this association. The deletion of employee *x*, however, is allowed since the destruction of an association link with his/her doctor does not violate the * multiplicity. A doctor is implicitly deleted only if the doctor is no longer recorded for any employee. Similar semantics apply to the explicit destruction of the link between an employee and a doctor, where the X' aspect of the ' binding is operative. In a sense, the ' binding provides for the garbage collection of doctors! (A doctor can never be explicitly deleted if recorded for any employee because of the |- binding.)

It is unclear how the semantics of this type of relationship can be described by the other declarative schemes discussed in this section.

5.2 Schemes for the ER model

ORN may also be compared to extensions to the ER model that have been suggested by others in order to specify or enforce association semantics, or *structural integrity constraints*. These extensions, however, are less declarative and more procedural in nature. Thus, while they are cited and briefly described below, no detailed comparisons are made with ORN.

Lazarevic and Misic (1991) extend entity and relationship types with constraints concerning actions that can or should be associated with different update operations. The actions indicate whether an update is to be restricted or should invoke more updates to enforce cardinalities. The constraints are intended to be used for the algorithmic design of integrity preserving update procedures.

Bouzeghoub and Metais (1991) allow the specification of more generalized integrity constraints and behavioral rules on the ER model. The behavioral rules provide for actions to be specified in the conclusion part that can enforce integrity rules. The constraints and rules are translated into enforcement constraint methods within an object database schema.

Finally, Balaban and Shoval (2000) build on these ideas by extending an enhanced ER model with *structure methods* to specifically enforce cardinality constraints. Templates for various structure methods are given that are built on top of *primitive update methods* that perform all update transactions. A user specifies association semantics by deciding among various processing options at certain points within the template code, e.g., reject an update or propagate an insert or delete to enforce a cardinality constraint.

5.3 Schemes in ODBMSs

Though limited, ODBMSs have implemented declarative capabilities for describing association semantics. Numerous commercial ODBMSs have been developed. They include CACHÉ (InterSystems Corp.), db4o (db4objects, Inc.), GemStone (GemStone Systems), O_2 (O_2 Technology), Objectivity/DB (Objectivity, Inc.), ObjectStore (Progress Software), ONTOS (Ontos, Inc.), Perst (McObject), and Versant Object Database (Versant Corporation). This section first discusses what declarative association semantics these systems generally provide and the equivalent

ORN capabilities. It then points out the importance of some ORN capabilities not provided by these systems.

What is generally provided by existing ODBMSs and reflected in the ODMG 3.0 object database standard (Catel 2000) is support for one-to-one, one-to-many, and many-to-many associations; referential integrity; and the automatic maintenance of inverse attributes. In terms of ORN, this is equivalent to supporting <0..1-to-0..1>, <0..1-to-*>, and <*-to-*> associations. Support for referential integrity means that on object deletion, all references to that object are automatically set to null or removed from object-based attributes, i.e., *relationship traversal paths* in ODMG 3.0. This is equivalent to the DELETING NULLIFIES referential integrity rule of Date (1981) and the default implicit destructibility bindings in ORN. The automatic maintenance of inverse attributes means that when changes are made to an object-based attribute, the system automatically updates the inverse attribute to reflect the change. This referential action is required to maintain the integrity of the representation of associations in an object database and has nothing to do with relationship semantics.

Some of the ODBMSs cited above, though not the ODMG 3.0 standard, also support a simple delete propagation capability. This is specified as an option for an object-based attribute, e.g. the macro parameter os_propagate_delete on the macro that defines an object-based attribute in Object Store. The semantics of simple delete propagation are that all objects referenced by the object-based attribute of an object are automatically deleted when the object is deleted. Delete propagation may cascade so that, for example, a whole part assembly is deleted when the main part is deleted. Delete propagation was added to ODBMSs to better support composite objects, i.e., the "is a part of" relationship. A variety of such relationships can be specified in ORN by employing any of the cascade bindings: ~?, ~!, or ~', where the ~ symbol is optional. The binding equivalent to the simple delete propagation available in some ODBMSs is the !.

Fig. 5.5 illustrates the use of the ! binding in a !<0..1-to-*> association between parts. Deleting a Part object results in the automatic deletion of the entire assembly, i.e., subtree, of parts having the explicitly deleted part object as its root.

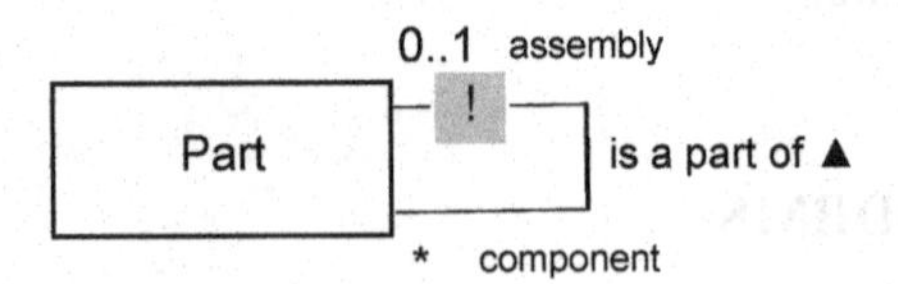

Fig. 5.5 Class diagram for association between parts

Support for the more precise multiplicity constraints of class diagrams and ORN is not available in most ODBMSs. These constraints are important in maintaining database integrity. Without them, integrity is at the mercy of code that is developed and maintained by database programmers. For example, in an object database, a <1-to-*> specification for the "works for" association, would (if it could be given) ensure that an employee always belongs to a unit and would cause an exception when an attempt is made to 1) delete a Unit object that references an Employee object, 2) set the unit object-based attribute of an Employee object to null, or 3) remove an Employee object reference from the employees object-based attribute of a Unit object. Without ODBMS support for the simple 1 constraint, inappropriately written or patched code can create invalid objects or even worse create inadvertent "logical garbage," e.g., employee objects that are no longer tied to any unit. Again, the unit and employees relationship is used here for illustration even though the inadvertent "logical garbage" problem may be more germane to other one-to-many relationships, e.g., a room set adrift that logically should not exist outside of a building. A "physical garbage" collection process as implemented by some object languages and ODBMSs can eliminate such logical garbage but does not solve the problem and in fact may exacerbate it. For example, a room erroneously set adrift is then deleted by the system!

Support for more precise multiplicities has additional benefits. It ensures that complex and composite objects can be defined precisely to the ODBMS in terms of their required number of constituents. This information is gleaned from the early stages of systems analysis and design and should not be lost or buried in code. A car pool has two or more, 2..*, riders and an automobile 2..5 doors. Also, precise multiplicities can be used to provide a conditional delete propagation.

This type of propagation is more desirable than the simple delete propagation offered by existing ODBMSs. A car pool can be implicitly deleted when its second to last rider is deleted. Or, in the analysis of detector data on a nuclear collision, a particle track can be automatically deleted when its last defining subtrack through a detector layer is deleted. Simple delete propagation does not permit objects to be deleted when they are no longer referenced by a prerequisite number of objects.

It is noteworthy that to my knowledge none of the declarative schemes for object models proposed in the late eighties and early nineties, e.g., Albano et al. (1991), have been adopted in commercial ODBMSs, and little work in this area has occurred since then. This is regrettable since integrating a more powerful declarative scheme like ORN into existing ODBMSs would dramatically improve software development productivity. The complex code required to implement association semantics, which are often quite subtle, would not have to be developed over and over again and maintained by ODBMS users.

5.4 Whole-Part Properties and Dimensions for Class Diagrams

Much effort has been devoted to understanding the "is a part of" relationships that often exist between objects. These relationships are often referred to as *whole-part* relationships. The aggregation and composition associations that can be defined in UML class diagrams are whole-part relationships. There has been much discussion and disagreement about what semantics, i.e., characteristics or properties, distinguish whole-part associations from other associations, what properties categorize whole-part associations into different types, and even what these properties mean (e.g., Albert et al. 2003, Barbier et al. 2003, Civello 1993, Henderson-Sellers and Barbier 1999, Odell 1994, Saksena et al. 1998, Snoeck and Dedene 2001, Winston et al. 1987). Some have argued that the UML specifications are ambiguous in defining aggregate and composition association types (e.g., Barbier et al. 2003, Henderson-Sellers and Barbier 1999). While declarative schemes have not been proposed for many of the semantics ascribed to whole-part associations, Albert et al. (2003) proposes some declarative properties that can be specified in class diagrams to supplement the white and black diamonds and better facilitate the generation of object-oriented code tailored to implement the modeled associations.

In this section we briefly examine some of the many properties that have been attributed to whole-part associations, many of which are applicable to all types of associations. The discussion for each property will:

- list within parentheses some of the different terms that have been used to describe it or its inverse;
- define its meaning and values in as precise yet uncontroversial manner as is possible;
- indicate whether a specific property or property value has been proposed as a *primary characteristic* for whole-part relationships (Saksena et al. 1998, Henderson-Sellers and Barbier 1999), which means that it along with other primary characteristics distinguish a whole-part association from other associations;
- describe its relevancy to database implementation,
- indicate how the semantic described by the property relates to ORN and provide an equivalent ORN representation if possible, and
- perhaps raise some issues.

More detail on each property is available in the above referenced papers.

5.4.1 Proposed association dimensions for class diagrams

I begin with the properties proposed in Albert et al. (2003), which are called association *dimensions*. For these properties, I give the permissible values using the proposed UML stereotype notation and give the required and default values intended to better distinguish aggregation, composition, and normal associations in UML class diagrams.

Multiplicity is the same as that defined by UML and ORN. In Albert et al. (2003), the multiplicity defaults to and must be 1..1 for the composite end of a composition. Interestingly, this requirement eliminates the possibility of a component ever existing independent of a composite, which UML allows. There is no default multiplicity for association ends that are not composite ends.

Reflexivity (inverse *irreflexivity*) indicates for an association whether an object can be linked to itself, i.e., whether object *a* "relates" to object *a* is allowed. The value «Reflexive» means it is allowed, and «Nonreflexive» means it is not. These stereotypes are placed in a class diagram near an association name. The value defaults to and must be «Nonreflexive» for shared aggregation and composition, and the default value is «Reflexive» for all other associations. Irreflexivity, i.e., a reflexivity of nonreflexive, is a commonly agreed upon primary characteristic of a whole-part association. Irreflexivity imposes a database constraint. For associations between objects of the same type (i.e., object instances of the same class), the DBMS must check at runtime to ensure that an object is not linked to itself. For other associations, type checking in the DML or programming language usually ensures such linkage is not possible and so reflexivity is really not an issue. The reflexivity semantic is orthogonal to and thus does not conflict with ORN semantics, which are relevant only to multiplicities and the destructibility of links.

Antisymmetry indicates for an association whether object *b* "relates to" object *a* is allowed when *a* "relates to" *b*. «Not Antisymmetric» means it is allowed, and «Antisymmetric» means it is not. These stereotypes are placed near an association name. The value defaults to and must be «Antisymmetric» for shared aggregation and composition, and the default for all other associations is «Not Antisymmetric». Antisymmetry is another commonly agreed upon primary characteristic of a whole-part association. Antisymmetry imposes a database constraint. For associations between objects of the same type and when link representation is supportive (which it is often not), the DBMS must check at runtime to ensure that *b* is not related to *a* when *a* relates to *b*. For other associations, type checking usually ensures this is not possible and so antisymmetry is really not an issue. The antisymmetry semantic is orthogonal to ORN semantics.

Visibility (inverse *encapsulation*) indicates for an association end whether objects can be accessed independent of (i.e., are not encapsulated in) related objects. The value «Visible» means objects can be accessed independent of related objects, and «Not Visible» means they cannot. The stereotypes are placed near an association end. The value defaults to and must be «Visible» for the composite end of a composition and «Not Visible» for the component end. The default value is «Visible» for all other association ends. (It is problematic that the value must be «Visible» for a composite since a composite may be a component in another composition.) This property affects how the association is represented in a database and accessed in the DML. The visibility semantic is orthogonal to ORN semantics.

Identity Projection (*exclusiveness*, *nonshareability*, inverse *shareability*) indicates for an association end whether an object at that end projects its identity onto its related object (i.e., whether the related object cannot be shared with another object). The value «Projected» means an object projects its identity (i.e., a related object cannot be shared), and «Not Projected» means the opposite. The stereotypes are placed near an association end. The value defaults to and must be «Projected» for the composite end of a composition and «Not Projected» for the component end. (Based on UML specifications, a white diamond implies «Not Projected» for the aggregate end of a shared aggregation, i.e., implies that components are shareable; however, Albert et al. (2003) apparently allows "shared aggregate" components to be specified as nonshareable.) As defined in Albert et al. (2003), identity projection for an association end imposes a database constraint, which in ORN can be specified by an upper bound multiplicity of 1 for that end; however, this identity projection does not include the "global exclusiveness" that is required by UML composition (OMG 2005, Barbier et al. 2003). Global exclusiveness means that not only can a component object not be shared by objects of a specific composite class, it also cannot be shared by objects of any related composite class. Such exclusiveness imposes an additional database constraint, which cannot be specified by multiplicities.

Temporal Behavior indicates for an association end whether objects at that end can be dynamically linked to or delinked from related objects. The value «Dynamic» means that they can throughout the whole life of an object, and the value «Static» means that such linking and delinking can only occur during the creation of an object. The stereotypes are placed near an association end. The value defaults to and must be «Static» for the component end of a composition, which forces the creation and linkage of components to be done via the composite class. The default value is «Dynamic» for all other association ends. A value of «Static» seems to impose a database constraint that disallows the creation and destruction of links by the DBMS once an object is created; however, Albert et al. (2003) does not make clear whether link destruction is altogether disallowed when one end of an association is specified as «Static» and the other end is specified as «Dynamic». Temporal behavior specifications may have more to do with which class controls link creation and destruction, i.e., the *ownership* of objects, in an object-oriented implementation rather than whether or not these operations are disallowed.

Equivalent semantics likely cannot be specified by ORN, although somewhat similar semantics are available. For an association defined between a class (or role) *S* and a related class (or role) (see Table 2.1), a 1-to-... association means that a related object, e.g., a component, must be linked to an *S* object within the transaction in which the related object is created. A X-<1-to-... association means that once a link is created, it cannot be explicitly destroyed, although the *S* object in the link can be exchanged with another. A X-<1-to-...>X- association means that once a link is created, it cannot be explicitly destroyed or changed in any way.

Delete Propagation indicates for an association end what actions, regarding links to related objects, are required when an object at that end is deleted. The value «Restrictive» means an object cannot be deleted if it has links to related objects, «Cascade» means that existing links and related objects must be deleted, and «Link» means that only existing links are deleted (but related objects remain). The stereotypes are placed near an association end. The value defaults to and must be «Cascade» for the composite end of a composition. The default value is «Link» for all other association ends. The delete propagation property defines referential actions that must be done by the DBMS. For a class (or role) *S* in an association with a related class (or role) (see Table 2.1), a «Restrictive» value given for delete propagation is equivalent to a |-<m-to-... association in ORN where m is the multiplicity for *S*. A «Cascade» value is equivalent to a |!<m-to-... association, and a «Link» value is equivalent to a <0..m_U -to-... association where m_U is some upper bound multiplicity. «Link» cannot be specified for an association end if the lower bound multiplicity at that end is not 0.

This completes our examination of the association properties and notation proposed by Albert et al. (2003). Now, we turn our attention to some of the other properties that have been discussed by others in addressing aggregation, composition, and whole-part relationships in general.

5.4.2 Primary characteristics for whole-part relationships

We begin with the properties that help set whole-part relationships apart from normal associations. Hypothetically, if all primary characteristics are true for an association, then the association is a whole-part association.

The *whole-part* property indicates for an association that it is a binary association where one end of the association plays the role of a whole and the other the role of a part (Barbier et al. 2003). Like all of the subsequent properties we examine, whole-part is itself a characteristic in that its value is true or false. Of course, whole-part can be specified in a class diagram by representing an association as a binary association and properly placing role names whole and part at the opposite ends of the association. Alternatively, one could just place a diamond at the whole end of the association, although Barbier et al. (2003) argues that aggregation as defined in UML does not encompass all types of whole-part relationships. Obviously, whole-part is a primary characteristic of a whole-part association. Independent of the other primary characteristics, it provides the semantic that objects playing the role of part are somehow "connected" to objects playing the role of whole to form an "aggregate whole." This semantic cannot be enforced by the DBMS and is orthogonal to ORN semantics.

An *emergent property* characteristic means for an association that each whole object has a property that emerges from its parts yet is independent of the properties of its parts. All whole-part relationships must have at least one emergent property according to Kilov and Ross (1994), Saksena et al. 1998, and Henderson-Sellers and Barbier (1999). For example, an emergent property of a bicycle is that it can be ridden. This property stems from the assembly of its parts yet cannot be deduced or derived from the properties of its individual parts. This semantic cannot be enforced by the DBMS and is orthogonal to ORN semantics. It may likely be that there is no reason to store an emergent property in the database.

A *resultant property* characteristic means for an association that each whole object has a property that results from its parts and is dependent on the properties of at least a subset of its parts and can be derived from these parts. All whole-part relationships must have at least one resultant property according to Kilov and Ross (1994), Saksena et al. 1998, and Henderson-Sellers and Barbier (1999). For example, a resultant property of a bicycle is its weight, which can be derived from the total weight of each of its parts. This semantic can be somewhat enforced by the DBMS if it requires the database designer to define a derived attribute for each whole whose derivation requires access to some of its parts; however, again it may be that there is no interest in storing such an attribute in the database. This semantic is orthogonal to ORN semantics.

Asymmetry (*Nonsymmetry*) means for an association both irreflexivity and antisymmetry. That is, when an association is asymmetric, or nonsymmetric, it is both «Nonreflexive» and «Antisymmetric» in terms of Albert et al. (2003) and as defined in the previous section. Asymmetry at the object level serves as a primary characteristic for whole-part relationships encompassing both irreflexivity and antisymmetry according to Kilov and Ross (1994), Saksena et al. 1998, and Henderson-Sellers and Barbier (1999).

According to this latter reference, antisymmetry at the type level is also a primary characteristic. At this level, class A objects "can be defined to be part of" class B objects and class B objects "can be defined to be part of" class A objects only if A and B are the same class, which is allowed. Whole-part relationships are often defined as intra-class associations, which means irreflexivity and thus asymmetry is not applicable at the type level.

Transitivity means for an association that if object *a* "relates to" object *b* and *b* "relates to" *c*, then *a* "relates to" *c*. According to UML, shared aggregation and composition are characterized by transitivity; however, transitivity is only considered a primary characteristic of a whole-part association by some (e.g., Saksena et al. 1998) but not by others (e.g., Barbier et al. 2003). When transitivity is attributed to a specific association, it is really only applicable for associations between objects of the same type. So why does UML attribute this to all aggregations? Transitivity imposes no database constraint, and thus the semantic cannot be checked by the DBMS. The semantic is orthogonal to ORN semantics.

5.4.3 Secondary characteristics for whole-part relationships

Finally, we examine proposed secondary characteristics for whole-part relationships that have not already been examined in Section 5.4.1. *Secondary characteristics* distinguish the different types of whole-part relationships and have been used to define taxonomies of such relationships (e.g., Winston et al. 1987, Odell 1994).

Configurationality (*functionality*) means for a whole-part association that there is a structural or functional relationship among the parts and the whole. For example, there is both a structural and functional relationship among the parts in a bicycle. On the other hand, there is no such relationship among the playing cards in a deck, and thus this whole-part relationship is *nonconfigurational*. This semantic cannot be enforced by the DBMS unless it would require some structure for or association between the part objects to be defined. The semantic is orthogonal to ORN.

Homeomerousity means for a whole-part association that the parts are the same "type of thing" as the whole. For example, a piece of cake is still cake. This semantic can be enforced by the DBMS by requiring that the whole and part objects be the same type. The semantic is orthogonal to ORN.

Existential dependency indicates for an association end that an object at that end is dependent on one or more related objects for its existence. Existential dependency is closely related to delete propagation; however, existential dependency merely means that an object *a*'s existence is dependent on its relation to an object *b* (e.g., Barbier et al. 2003) without indicating what should be done if *b* is deleted. This dependency is simply represented by ensuring that the lower bound multiplicity is greater than or equal to 1 at the *b* end of the association. This means that existential dependency as a property is redundant and is simply a multiplicity constraint that the DBMS should enforce. In ORN, this constraint is given as an ...-to-$m_L..m_U$ association between *a* and *b* where $m_L \geq 1$, although $m_L..m_U$ is usually 1..1.

Now, if *a*'s existence is dependent on a relationship with *b*, what should be done when *b* is deleted? This is where delete propagation comes into play. A «Restrictive» value for delete propagation in Albert et al. (2003) would not allow the deletion of *b* and a «Cascade» value allows the deletion of *b* but implicitly deletes *a*. A «Link» value allows the deletion of *b*, destroys its link with *a*, and allows *a* to remain, and thus this value is incompatible with existential dependency. We have already seen the ORN equivalencies for these values.

We should note that Albert et al. (2003) does not provide for the case where an object in an association can be related to multiple objects but is dependent on being related to a minimum number for its existence—i.e., for a ...-to-$m_L..m_U$ association, $m_L \geq 1$ and m_L.>.m_U. For example, a family must have at least two people but can have more. If we do not want to allow the deletion of a person if the family's existence is dependent on this person, i.e., a "«Restrictive»-like propagation," then the proper ORN for the relationship between families and persons is <...-to-2..*>. If we want the family deleted when membership falls below two because of the deletion of a person, i.e., a "«Cascade»-like propagation," then the proper ORN is <...-to-2..*>|?.

Separability, *mutability* (*changeability*, *variance*, inverse *immutability* and *invariance*), and *lifetimes* are properties that are closely related. They are also related to temporal behavior as defined in Albert et al. (2003) and existential dependency. Exactly how these properties relate to each other and the clarity of their use in UML specifications has been much discussed. These issues are not debated here and no position is taken. Briefly, separability means that parts can be separated from the whole and mutability means that the relationships between wholes and parts can be changed. These properties have been applied to both associations and association ends. The lifetimes property indicates how the lifetimes of the whole and its parts overlap. As previously shown, the X- binding in ORN at one or both ends of an association specifies the destructibility of association instances, i.e., separability, and the changeability of association instances once created. Other bindings specify whether related objects must live or die when attempts are made to kill objects or destroy the links between them.

5.5 Conclusion

Based on the comparisons given in this chapter and the examples given in the previous chapter, ORN can be seen as a practical declarative scheme that facilitates the implementation of a large variety of associations. The notation makes expedient use of multiplicities and a limited number of binding symbols to minimize syntax, semantics, constraint inconsistencies, and the number and complexity of concepts that a database developer must confront. In addition, ORN has the following features.

- It is defined at both a conceptual and logical level, meaning it can be incorporated into ER and class diagrams and it can be adapted to the relational and object models and DDLs (which will become more apparent in later chapters).
- It is based on and consistent with association multiplicities, a well-understood concept.
- It provides the basis for defining the extent of complex and composite objects.
- It incorporates the equivalent of the referential integrity rules first proposed for relational databases as well as null, foreign key, and surjectivity constraints.
- It provides for the automatic garbage collection of objects.
- It provides a means to specify most of those properties associated with whole-part relationships that can be enforced by a DBMS.
- It is orthogonal to the other properties, meaning that while these properties cannot be expressed by ORN, their specifications are not incompatible and thus can be given with ORN. (Most of these properties, however, while theoretically interesting, have little to do with the actual implementation of whole-part relationships in a database.)

The purpose of this chapter was to permit all of the above claims for ORN to be better assessed.

Part II

Using ORN to Develop a Database System

Chapter 6

ORN Additive

A Tool for Extending SQL Server with ORN

ORN Additive for Transact-SQL (T-SQL) adds ORN to Microsoft SQL Server (Microsoft 2005). The "additive" increases the productivity of developing database systems and improves data integrity in these systems. With ORN Additive, implementation is easier as database models more directly map to the required DDL statements of T-SQL. More specifically, relationships can be defined in .sql files with the multiplicities and bindings given in UML class diagrams. These multiplicities and bindings are then automatically enforced by the database system as database changes are made using an ORN-extended T-SQL. ORN Additive was developed to show the feasibility of defining associations to an RDBMS using ORN. SQL Server was chosen because T-SQL provides INSTEAD OF triggers and transaction processing operations within these triggers. These capabilities are crucial to implementing ORN.

Although ORN Additive was developed as a prototype, it can be a very productive tool for developing many relational database applications. Many lines of T-SQL code, involving complex triggers and stored procedures, are automatically generated by the tool. This code would otherwise have to be painstakingly implemented, tested, and maintained by database developers. The complete ORN Additive tool, including the *ORN Additive for Transact-SQL User's Guide*, can be downloaded from the author's website at the URL given in the Preface. The user's guide provides all of the technical details for installing and using the tool.

This chapter provides an overview of ORN Additive, first discussed in Ehlmann (2007). It is partially based on this paper[1] and the ORN Additive user's guide. Section 6.1 examines ORN Additive capabilities, operation, and architecture. Section 6.2 examines ORN Additive DDL statements, which extend the DDL type statements of T-SQL. Section 6.3 examines ORN Additive DML statements, which extend the DML type statements of T-SQL. Section 6.4 concludes by summarizing the benefits and limitations of the modeling and implementation approach made possible by ORN Additive and by indicating how the limitations can be overcome.

6.1 Capabilities, Operation, and Architecture

The ORN Additive tool consists of a DDL utility and a DML utility. This section briefly describes how these utilities essentially add ORN to SQL Server.

---

[1] Portions reprinted, with permission, from "ORN Additive: shrinking the gap between database modeling and implementation," *Proc ICIS Conf,* IEEE Computer Society, 555-560. © 2007 IEEE.

B.K. Ehlmann, *Object Relationship Notation (ORN) for Database Applications*,
Advances in Database Systems 39, DOI 10.1007/978-0-387-09554-7_6,

6.1.1 Capabilities

The DDL utility is a postprocessor that processes ORN Additive DDL statements, i.e., --+<> statements, which supplement DDL-type T-SQL statements that define a database. (The + indicates addition and the <> recalls the diamonds in an ER diagram and the way relationships have been traditionally defined in data models.) The --+<> statements may be interspersed as comments within T-SQL statements (see Fig. 6.1) and are processed by the DDL utility after the T-SQL statements have been executed. What results is a defined *ORN-enhanced database*, i.e., one where *<association>*s have been given for foreign keys.

```
USE CompanyDB;
--+<> USE CompanyDB;
CREATE TABLE Carpool (
  id          VARCHAR(8) PRIMARY KEY,
  ...
);
CREATE TABLE Employee (
  ssn         CHAR(11) PRIMARY KEY,
  ...
  carpoolid   VARCHAR(8) CONSTRAINT belongsTo REFERENCES Carpool(id),
--+<> belongsTo ?<2..15-TO-0..1>;
);
GO
```

Fig. 6.1 Example of T-SQL query file including two ORN Additive DDL statements

The DML utility is a preprocessor that processes ORN Additive DML statements, i.e., +<> statements. These statements extend DML-type T-SQL statements that specify the USE of an ORN-enhanced database and provide explicit transaction processing within queries that modify the database. The +<> statements are interspersed with T-SQL statements as needed (see Fig. 6.2) and are processed and translated into T-SQL statements before the query file is executed.

```
+<> USE CompanyDB;
+<> BEGIN TRAN;
  DELETE Employee WHERE ssn = '555-55-5555';  -- Results in deletion of a row
                                              -- in Carpool if lowerbound 2 is violated.
+<> COMMIT TRAN;
GO
```

Fig. 6.2 Example of T-SQL query file including three ORN Additive DML statements

The DDL utility generates T-SQL statements based on the given ORN Additive statements and the metadata views that define the database. The generated T-SQL statements when executed add objects to the database, e.g., triggers, which support the enhanced capabilities provided by the ORN-defined relationships. These capabilities become apparent when queries are executed that affect relationships—i.e., that insert rows, delete rows, and update primary and foreign keys.

6.1.2 Operation and architecture

The steps in using ORN Additive to develop a database system are given below. These steps can be seen in Fig. 6.3, which shows the architecture and data flow for the use of ORN Additive with SQL Server.

1. Develop and execute the required DDL-type T-SQL statements to define the database (or alter an existing database). For tables that are to be related by relationships defined by ORN, i.e., *ORN-related tables*, ensure that:

   - All foreign keys in the table are defined by named constraints.
   - No ON DELETE or ON UPDATE clause is given for these constraints.
   - No NOT NULL constraint is given for any foreign key column.

2. Once the database has been defined to SQL Server, develop a query file containing ORN Additive statements, i.e., --+<> statements, that provide the ORN, e.g., <*-to-1>, for the foreign key constraints. Process this query file using the ORN Additive DDL utility (+ornddl). (Note that since the ORN Additive statements are "structured comments," they can be appropriately interspersed within, i.e., added to, a .sql file containing the T-SQL statements developed in Step 1.) When no errors are found by the DDL utility, three query files are generated:

   - an *ORN primer file*, ORN_Primer.sql
   - an *ORN additive file*, ORN_AdditiveFor_*database_name*.sql
   - an *ORN header file*, ORN_HeaderFor_*database_name*.sql

3. Execute the ORN primer file to "prime" SQL Server for ORN Additive. Executing this file defines ORN Additive error messages to SQL Server and stores ORN system stored procedures in a database, which is referenced as the "ORN sp Database" in Fig. 6.3 and is the master database by default. This step need only be done once for an SQL Server instance.
4. Execute the ORN additive file to provide for subsequent database operations based on your *ORN-defined foreign keys*. Executing this file adds metadata as well as a number of triggers and stored procedures to the database.
5. When developing T-SQL queries that Insert rows into, Delete rows from, and/or Update rows in ORN-related tables, i.e., *IDU queries*, do one of the following:

   a) If the IDU query is a *simple IDU query*, i.e., it doesn't use explicit transactions on ORN-related tables and uses just one ORN-enhanced database, copy & paste the ORN header file (from Step 2) to the beginning of the IDU query.
   b) If the IDU query is not a simple IDU query, use appropriate ORN Additive statements (+<> USE database, +<> BEGIN TRAN, +<> SAVE TRAN, +<> COMMIT TRAN, and +<>ROLLBACK TRAN) in place of the corresponding T-SQL statements. Save these IDU query files as .sql+ files and process them using the ORN Additive DML utility (+orndml). This utility creates a .sql file when no errors are encountered. This .sql file is the same as the .sql+ file except that the ORN header is included and ORN Additive

statements are replaced by their corresponding T-SQL statements along with T-SQL statements that EXECute some of the ORN system stored procedures created in Step 3.

6. Execute the .sql files created by Step 5 to make changes to the database. ORN semantics will be automatically enforced by the database system.

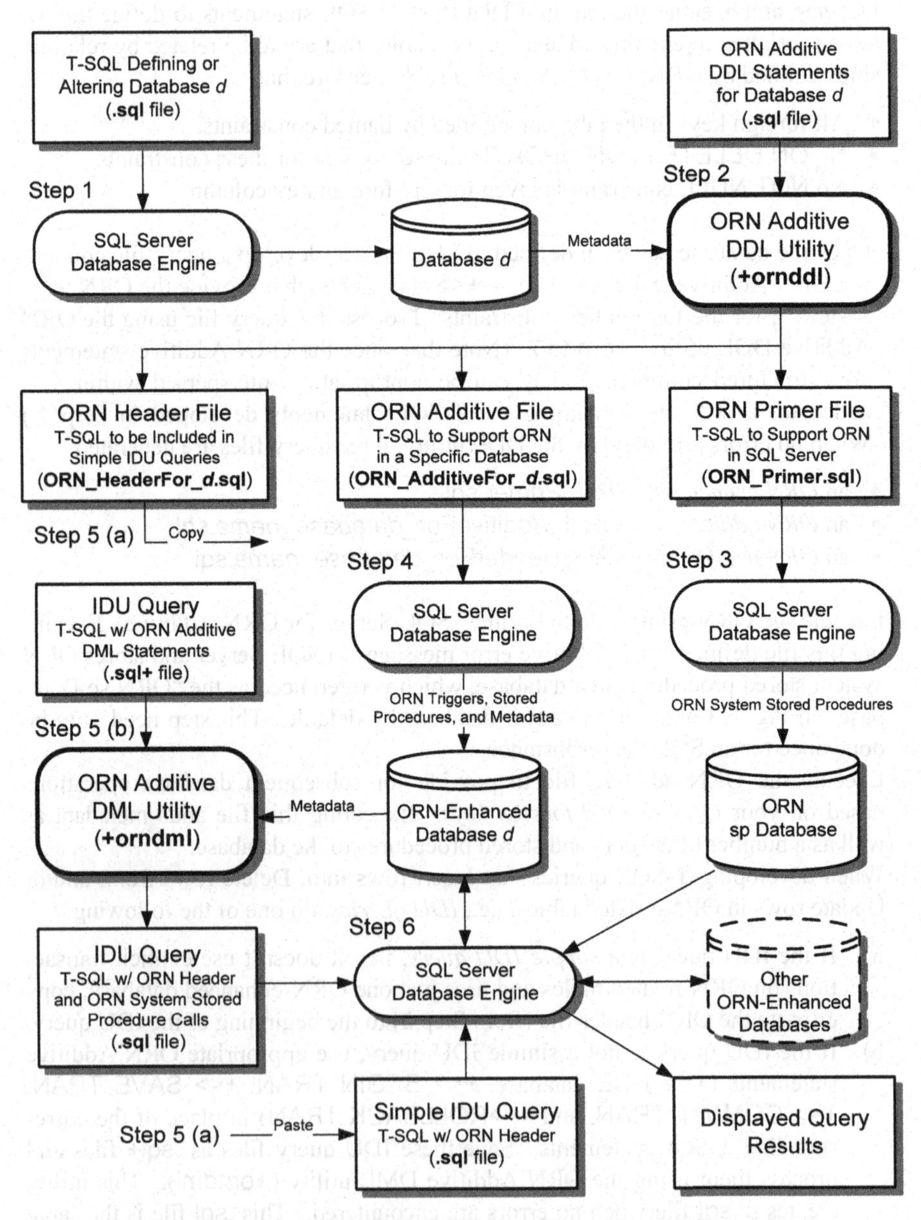

Fig. 6.3 Architecture and data flow for the use of ORN Additive with SQL Server

6.1.3 The **+ornddl** *command*

The +ornddl command executes the ORN Additive DDL (+ornddl) utility. This utility "postprocesses" ORN Additive DDL statements, i.e., statements beginning with --+<>, in order to add ORN notation to the referential constraints defined for an SQL Server database. The input file is a query, i.e., .sql, file that usually also contains T-SQL DDL-type statements that have been previously processed to create the database. When +ornddl finds no errors, the ORN primer file, ORN additive file, and ORN header file are generated in the directory of the input query file.

Table 6.1 summarizes the options that can be given in the +ornddl command. More detail, including all option defaults, are given in the ORN Additive user's guide. Some examples of the +ornddl command are given below.

```
+ornddl –U MyUserId –P MyPassWrd
+ornddl –i create_tables.sql –np –dGB –sp globalSysProcsDB
+ornddl –U smithJB –P snoppy02 –S coServer\sys3 –i defineDB.sql -so
```

Table 6.1 +ornddl Command Options

Option	Description
-U	Allows a user login id to be given in order to connect to the SQL Server. By default Microsoft Windows Authentication mode is used.
-P	Allows a user password to be given.
-E	Specifies that a trusted connection be used to log in to the SQL Server.
-S	Specifies the instance of SQL Server to which to connect.
-i	Identifies the input query file to be processed.
-so	Specifies syntax only, meaning that no output files will be generated.
-np	Specifies that no ORN primer file be generated.
-dGC	Specifies debug Generated Code, meaning that Print statements will be included in the ORN Additive file to allow tracing of generated code execution.
-sp	Specifies the name of the sp database that will be used to store ORN system generated stored procedures. By default this is the master database.
-v	Specifies that the version number of +ornddl be displayed.
-?	Specifies that the syntax of the +ornddl command be displayed.

6.1.4 The **+orndml** *command*

The +orndml command executes the ORN Additive DML (+orndml) utility. This utility preprocesses ORN Additive DML statements, i.e., statements beginning with +<>, that are interspersed between T-SQL statements in a query file. These ORN Additive statements are required to enforce the semantics of ORN-defined foreign keys. More specifically, these statements must be used in IDU queries when explicit transactions are used to insert into, delete rows from, or update rows within ORN-related tables. They must also be used when ORN-related tables from multiple ORN-enhanced databases are being changed within a transaction.

Table 6.2 summarizes the options that can be given in the +orndml command. More detail, including all option defaults, are given in the ORN Additive user's guide. Some examples of the +orndml command are:

```
+orndml –U MyUserId –P MyPassWrd
+orndml –i updOrder.sql+ –sp globalSysProcsDB
+orndml –U smithJB –P snoppy02 –S coServer\sys3 –o test2.sql -so
```

Table 6.2 +orndml Command Options

Option	Description
-U	Allows a user login id to be given in order to connect to the SQL Server. By default Microsoft Windows Authentication mode is used.
-P	Allows a user password to be given.
-E	Specifies that a trusted connection be used to log in to the SQL Server.
-S	Specifies the instance of SQL Server to which to connect.
-i	Identifies the input query file, normally a .sql+ file, to be processed.
-o	Identifies the output query file that is generated. The default, when the input file is given as a .sql+ file, is the same file name but with a .sql suffix.
-so	Specifies syntax only, meaning that no output file will be generated.
-sp	Specifies the name of the sp database that contains the ORN system generated stored procedures. By default this is the master database.
-v	Specifies that the version number of +orndml be displayed.
-?	Specifies that the syntax of the +orndml command be displayed.

6.2 ORN Additive DDL Statements

ORN Additive DDL statements specify the ORN for relationships defined in a database schema. Essentially, an *<association>* is provided for all foreign key constraints that implement relationships whose semantics are to be governed by ORN. The *<association>* transfers to the database system the multiplicities and bindings that have been specified for a relationship in an ORN-extended database model.

When modifications are made to a database schema that would add, remove, or change relationships, appropriate modifications must be made to the ORN Additive statements for the database and **+ornddl** must be rerun to process these statements. Also, if required, appropriate changes should be made to the database so that the current data is consistent with the new ORN specifications.

6.2.1 USE statement

The USE statement specifies the database to the +ornddl postprocessor providing the context for subsequent CONSTRAINT statements. An example of this statement is:

```
--+<> USE companyDB;
```

6.2.2 **CONSTRAINT** *statement*

The CONSTRAINT statement provides the ORN for a relationship in the database, the relationship being implemented as a foreign key, i.e., referential, constraint. The name of the constraint along with the applicable *<association>* are given in the statement. An ON UPDATE CASCADE clause can also be given to specify the cascading of a primary key value change in a referenced table row to matching foreign keys in referencing table rows. Some examples of this statement are:

```
--+<> CONSTRAINT belongsTo ?<2..15-TO-0..1>;
--+<> belongsTo ?<2..15-TO-0..1>;
--+<> CONSTRAINT InventorySchema.isPartOf <*-TO-0..1>!
--+<>      ON UPDATE CASCADE;
```

The named foreign key constraint cannot have any ON DELETE or ON UPDATE referential actions defined in the SQL other than NO ACTION. Delete referential actions are incorporated within ORN. Also, a foreign key column cannot have a NOT NULL constraint. This constraint is also implicitly incorporated into ORN. If a foreign key is part of the primary key, an explicit "link destruction" is not allowed.

The semantics for an *<association>* in the CONSTRAINT statement are those given in Chapter 2, except that in the context of SQL, "class" translates to "table," "object" to "row," and "link" to a row reference by a foreign key. Table 6.3 describes ORN semantics in terms of the relational model. In this model a link is destroyed when its implementing, row reference is set to NULL. A link is explicitly changed when a row reference is set to reference another row. In a relational database and thus in a CONSTRAINT statement, an *<association>* cannot be many-to-many, e.g., <*-to-1..*>. Such associations must, of course, be remodeled as two one-to-many associations before implementation.

6.2.3 **DELETE** *statement*

The DELETE statement provides for the archiving of information within a row before its deletion. The statement specifies an *application-defined deletion stored procedure* that is to be used to delete a row in a given ORN-related table. An example of this statement is:

```
--+<> DELETE Employee USING HRS.usp_ArchiveAndDelete_Emp;
```

The triggers generated by ORN-Additive execute the given procedure to delete a row in the given table instead of executing a DELETE statement. The procedure is executed for explicit and implicit deletion and provides for application-specific actions to occur before or after row deletion. It is passed one or more parameter values representing the primary key attribute values of the row to be deleted.

Table 6.3 ORN Semantics for the relational model

Semantics are given in terms of a subject table *S* with multiplicity *m* and binding *b* in an association *A* with a related table *R* (which could be table *S* in a different role).

S | b | m | A | R

An *A* link is implemented by a foreign key reference, *fk*, from a row of *S* to a row of *R* or vice versa. An *A* link, now an "*A* reference," is destroyed when an *fk* reference is set to NULL or changed to reference a new row.

<*multiplicity*>: Semantics are similar to those in UML. Essentially, *m* indicates a lower bound and upper bound on the number of rows of table *S* that reference via *fk* a particular row of table *R* (or the number of rows of table *R* that can be referenced via fk by a row of table *S*).

An *S* row can be created provided this does not violate *m*. The check for a lower bound violation is deferred until transaction commit. A *fk* can be set provided this does not violate *m*. The check for an upper bound violation is immediate. The enforcement of *m* on the deletion of an *S* row or on setting *fk* is determined by the binding *b*.

<*binding*>: A **|** in *b* denotes a "cut" and an Implicit, i.e., system initiated, destruction of an existing *A* reference that must occur on deletion of an *S* row. An **X** in *b* denotes a "cross out" and an eXplicit, i.e., user initiated, destruction of an *A* reference.[1]

```
CREATE TABLE S (
  pk ... PRIMARY KEY,
  fk ... CONSTRAINT A REFERNCES R(pk),
  --+<> A b<m-to-...>...;
  ...
);
CREATE TABLE R (
  pk ... PRIMARY KEY,
  ...
);
                or
CREATE TABLE S (
  pk ... PRIMARY KEY,
  ...
);
CREATE TABLE R (
  pk ... PRIMARY KEY,
  fk ... CONSTRAINT A REFERNCES S (pk),
  --+<> A <...-to- m>b ...;
  ...
);
```

An *S* row deletion and an explicit destruction of an *A* reference are *complex object operations*. Deletion of an *S* row succeeds only if all existing references involving that row are implicitly destructible. Also, deletion of an *S* row or destruction of an *A* reference succeeds only if all required implicit row deletions succeed.

<*di*>: A destructibility indicator in *b* specifies the destructibility of an *A* (i.e., *fk*) reference. The meaning of each indicator is given below. This meaning can alternatively be described by the actions taken on an attempt to destroy an *A* reference. These actions are given in brackets. If a <*di*> is given after a **|**, it applies to the implicit reference destruction; if given after an **X**, it applies to explicit reference destruction; and if given alone, it applies to both. If a <*di*> is not given, i.e., is nil, for implicit reference destruction, explicit reference destruction, or both, default destructibility applies to whichever.

nil — *Default destructibility*. A reference can be destroyed provided this does not violate *m*.[2] [Destroy the reference. If *m* is violated [2], raise an exception [3].]

- — *Negative destructibility*. A reference cannot be destroyed. [Raise an exception.[3]]

~? or **?** — *Conditional cascade destructibility*. A reference can be destroyed, but if this violates *m* (**?**), the destruction must be cascaded (~) to the related *R* row, i.e., this row must be implicitly deleted. [Destroy the reference. Delete the related *R* row**?** If *m* is violated, yes; else no.]

~! or **!** — *Emphatic cascade destructibility*. A reference can be destroyed, but the destruction must be cascaded (~) to the related *R* row. [Destroy the reference. Delete the related *R* row**!**]

~' or **'** — *Tentative (or qualified) cascade destructibility*. A reference can be destroyed, but an attempt must be made to cascade (~) the destruction to the related *R* row; however, this implicit *R* row deletion must be undone if it fails, but is required if and only if its undoing would violate *m*.[2] (Think of the **'** as a "pruned back !" or as a "qualifying footnote reference" on the cascade.) [Destroy the reference. Delete the related *R* row.**'** (**'** – If an exception occurs, undo the delete and then, if *m* is violated [2], raise an exception [3].)]

1 - An *fk* reference change done as a single operation that replaces an S row with another is not treated as an explicit link destruction relative to table *S* (but is relative to *R*) and is allowed if allowed by other multiplicities and bindings.
2 - The check for a lower bound violation is deferred until the end of the current complex object operation.
3 - The current complex object operation is undone.

6.2.4 SET ORN_MESSAGE_NUMBER_BASE *statement*

The SET ORN_MESSAGE_NUMBER_BASE statement allows one to change the numbering of ORN error messages so that they do not conflict with the numbering of other user-defined error messages. An example of this statement is:

```
--+<> SET ORN_MESSAGE_NUMBER_BASE 60000;
```

6.2.5 Example of a query file with ORN Additive DDL statements

Fig. 6.4 shows an example of a database model defined by an ORN-extended class diagram. (It is the same as the model given in Fig. 4.18.) Fig. 6.5 shows how ORN Additive DDL statements are interspersed within T-SQL statements to implement this database model. The actual process of mapping the database model to this "ORN-extended T-SQL" is discussed in Chapter 8.

6.3 ORN Additive DML Statements

ORN Additive DML statements extend certain T-SQL DML-type statements to enforce the ORN semantics for one or more ORN-enhanced databases. The statements identify the databases to be manipulated; extend transaction processing capabilities to facilitate implicit, recursive database processing for binding enforcement; and allow lower bound multiplicity constraints to be deferred until the transaction commit.

ORN Additive DML statements are only required in IDU queries that use explicit transactions to modify an ORN-enhanced database or use multiple ORN-enhanced databases. Explicit transactions are often required when inserting rows into the database as ORN lower bound multiplicities are enforced on the commit of a transaction, including the implicit transaction created for a single INSERT statement.

6.3.1 USE *statement*

The USE statement is like the USE statement of T-SQL and substitutes for it when updating an ORN-enhanced database. An example of a USE statement is:

```
+<> USE companyDB;
```

From +<> USE statements, the +orndml preprocessor determines the ORN-enhanced databases that can be modified by transactions in a T-SQL query file. For an ORN-enhanced database to be modified in a transaction, its name must appear in a +<> USE statement somewhere in the query file.

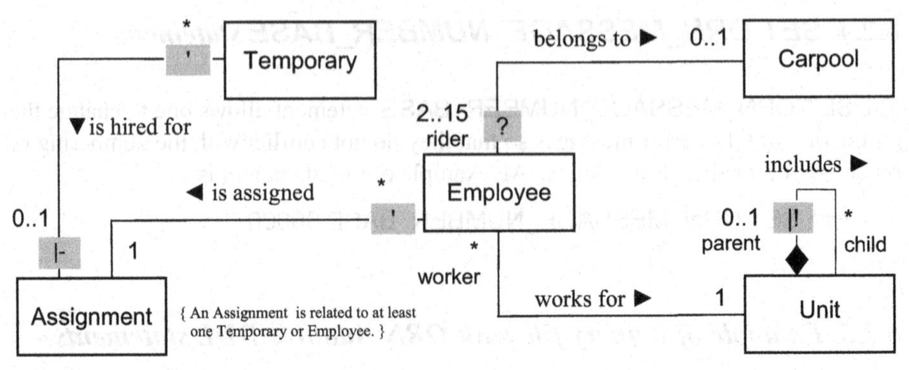

Fig. 6.4. Model for a company database

```
USE CompanyDB;
--+<> USE CompanyDB;
CREATE TABLE Unit (
  name        VARCHAR(20)  PRIMARY KEY,
  ...
);
CREATE TABLE Carpool (
  id              VARCHAR(8) PRIMARY KEY,
  ...
);
CREATE TABLE Assignment (
  code          CHAR(3) PRIMARY KEY,
  ...
);
CREATE TABLE Employee (
  ssn            CHAR(11) PRIMARY KEY,
  ...
  carpoolid     VARCHAR(8) CONSTRAINT belongsTo REFERENCES Carpool(id),
--+<> belongsTo ?<2..15-TO-0..1>;
  unitName     VARCHAR(20) CONSTRAINT worksFor REFERENCES Unit(name),
--+<> worksFor <*-TO-1>;
  assignCode  CHAR(3) CONSTRAINT isAssigned REFERENCES Assignment(code)
--+<> isAssigned '<*-TO-1> ON UPDATE CASCADE;
);
CREATE TABLE Temporary (
  ssn            CHAR(11) PRIMARY KEY,
  ...
  assignCode  VARCHAR(3) CONSTRAINT isHiredFor REFERENCES Assignment(code)
--+<> isHiredFor '<*-TO-0..1>|- ON UPDATE CASCADE;
);
ALTER TABLE Unit ADD
  parent          VARCHAR(20) CONSTRAINT includes REFERENCES Unit(name);
--+<> includes <*-to-0..1>|!;
GO
--+<> DELETE Employee ROW USING HRS.usp_ArchiveAndDelete_Emp;
--+<> SET ORN_MESSAGE_NUMBER_BASE 60000;
```

Fig. 6.5 T-SQL with ORN Additive statements implementing model in Fig. 6.4

6.3.2 ***BEGIN TRANSACTION*** *statement*

The BEGIN TRANSACTION statement is like the BEGIN TRANSACTION statement of T-SQL and substitutes for it when inserting, deleting, or updating rows in any ORN-related table. Two examples of this statement are:

```
+<> BEGIN TRAN;
+<> BEGIN TRANSACTION T1;
```

6.3.3 ***SAVE TRANSACTION*** *statement*

The SAVE TRANSACTION statement is like the SAVE TRANSACTION statement of T-SQL and substitutes for it when inserting, deleting, or updating rows in any ORN-related table. An example of this statement is:

```
+<> SAVE TRAN savePoint1;
```

6.3.4 ***COMMIT TRANSACTION*** *statement*

The COMMIT TRANSACTION statement is like the COMMIT TRANSACTION statement of T-SQL and substitutes for it when inserting, deleting, or updating rows in any ORN-related table. Two examples of this statement are:

```
+<> COMMIT TRAN;
+<> COMMIT TRANSACTION T1;
```

Unlike the T-SQL COMMIT statement, this statement checks *deferred constraints*, which for ORN are lower bound multiplicity constraints that may have been violated because of database changes made within the transaction. When violated, an error "Lower bound multiplicity was violated for ..." is raised. The error identifies the applicable foreign key constraint and the related table for which the multiplicity was violated. The violation must be corrected before the transaction can be committed, or the transaction must be rolled back.

6.3.5 ***ROLLBACK TRANSACTION*** *statement*

The ROLLBACK TRANSACTION statement is like the T-SQL ROLLBACK TRANSACTION statement and substitutes for it when inserting, deleting, or updating rows in any ORN-related table. Some examples of this statement are:

```
+<> ROLLBACK TRAN;
```

+<> ROLLBACK TRANSACTION T1;
+<> ROLLBACK TRAN savePoint1;

6.3.6 SET RXC_MODE statement

The SET RXC_MODE statement changes the Relationship eXChange mode for subsequent transactions. Examples of this statement are:

+<> SET RXC_MODE ON;
+<> SET RXC_MODE OFF;

When a transaction is in RXC mode, checks on lower bound multiplicities, which may invoke binding-related actions, e.g., implicit deletes, and which are often re-checked at the end of a complex operation, do not invoke such actions and are deferred until the COMMIT of the application-specified transaction. This permits "harmless" relationship exchanges within a transaction.

The RXC mode feature complements ORN. It was introduced in Chapter 3 and will now be discussed in more detail in the context of a relational database implementation. For this discussion and the examples to follow, assume the database as defined in Fig. 6.5 and the Employee, Carpool, and Unit tables given in Fig. 6.6.

Employee

ssn	...	carpoolid	org	...
111-11-1111		East	Accounting	...
222-22-2222		West	Personnel	...
333-33-3333		West	Accounting	
999-99-9999		East	Accounting	...

Carpool

id	...
East	...
West	...

Unit

name	...
Accounting	...
Personnel	...

Fig. 6.6 Sample tables for database defined in Fig. 6.5

Suppose we wish to exchange carpools for employees 222-22-222 and 999-99-9999. That is, carpools East and West have agreed to a trade. Without RXC mode set, the transaction below will fail because the first UPDATE would result in the implicit deletion of the Carpool row with id of West (because of the ? binding and lower bound multiplicity violation of 2 for Employee in the belongsTo constraint).

+<> BEGIN TRAN;
UPDATE Employee SET carpoolid = 'East' WHERE ssn = '222-22-2222';
UPDATE Employee SET carpoolid = 'West' WHERE ssn = '999-99-9999'
+<> COMMIT TRAN;

If, however, RXC mode is set as below, the transaction will succeed since no implicit deletion of carpool West is done. The lower bound multiplicity check to ensure that the carpool West has two employees is deferred until the +<> COMMIT TRAN, at which time the carpool will have its required two employees.

```
SET RXC_MODE ON;
+<> BEGIN TRAN;
UPDATE Employee SET carpoolid = 'East' WHERE ssn = '222-22-2222';
UPDATE Employee SET carpoolid = 'West' WHERE ssn = '999-99-9999'
+<> COMMIT TRAN;
SET RXC_MODE OFF;
```

If carpool East had fifteen riders before the above transaction, the transaction would fail. This is because the first UPDATE would result in an upper bound multiplicity violation. Setting RXC mode defers lower bound multiplicity checking but not upper bound checking. The transaction below, however, would succeed.

```
SET RXC_MODE ON;
+<> BEGIN TRAN;
UPDATE Employee SET carpoolid = 'West' WHERE ssn = '999-99-9999'
UPDATE Employee SET carpoolid = 'East' WHERE ssn = '222-22-2222';
+<> COMMIT TRAN;
SET RXC_MODE OFF;
```

RXC mode is not always required for relationship changes. If an UPDATE changes a foreign key that references a row to reference another row, then actions related to the explicit binding and multiplicity for the referenced table are not invoked. (See footnote 1 in Table 6.3, where *S* is the referenced table.) For example, RXC mode is not required for the relationship change shown below that transfers employee 111-11-1111 from the Accounting to the Personnel unit.

```
UPDATE Employee SET org = 'Personnel' WHERE ssn = '111-11-1111';
```

Despite the fact that this operation entails destroying the reference between the Employee row with ssn of 111-11-1111 and the Unit row with name of Accounting, the 1 multiplicity for the Unit table will not be violated in the end. The UPDATE is done within an implicit ORN-extended transaction.

Doing the above relationship change as a two-step process, however, as shown below, would result in a lower bound violation error if RXC_MODE was not ON.

```
+<> BEGIN TRAN;
UPDATE Employee SET org = NULL WHERE ssn = '111-11-1111'
UPDATE Employee SET org = 'Personnel' WHERE ssn = '111-11-1111';
+<> COMMIT TRAN;
```

6.3.7 ENABLE/DISABLE ORN_TRIGGERS *statement*

The ENABLE/DISABLE ORN_TRIGGERS statement enables or disables the INSERT, UPDATE, and DELETE triggers used to support ORN. By default, these triggers are enabled. Examples of this statement are:

```
+<> DISABLE ORN_TRIGGERS;
+<> ENABLE ORN_TRIGGERS;
```

Disabling ORN triggers means that the constraints and actions associated with ORN-defined foreign keys will not be enforced. This can result in a database being in an inconsistent state in regard to the multiplicity constraints given in the <*association*>s defined for one or more foreign keys. For efficiency reasons, it is safe to disable ORN triggers when doing any operation that does not violate any multiplicity constraint or invoke any non-default binding action, e.g., updating columns in ORN-related tables that are not columns in a primary key or ORN-defined foreign key.

6.3.8 Example of query file with ORN Additive DML statements

Fig. 6.7 shows a query (**.sql+**) file that includes ORN Additive DML statements. The query is based on the database defined in Fig. 6.5. Comments within the code indicate actions that may result from enforcing ORN semantics. The first comment makes sense when one considers that employee 444-44-4444 may be one of just two employees in the Bristol carpool. The second comment makes sense when one considers that employee 777-77-7777 may not exist in the database.

```
+<>USE CompanyDB;
GO
+<>BEGIN TRAN;
  INSERT INTO Carpool VALUES ('Airport', ...);
  UPDATE Employee
    SET carpoolId = 'Airport' WHERE ssn = '444-44-4444' OR ssn = '777-77-7777';
    -- Will implicitly delete any carpool if reassignment of an employee to the 'Airport' carpool
    --    results in less than two employees.
BEGIN TRY
+<> COMMIT TRAN;
    -- Will raise exception if two employees have not been assigned to the 'Airport' carpool.
  PRINT 'Transaction committed.';
END TRY
BEGIN CATCH
+<> ROLLBACK TRAN;
  Print 'Database transaction rolled back';
END CATCH
GO
```

Fig. 6.7 Query (.sql+) file with ORN Additive DML statements for database defined in Fig. 6.5

Fig. 6.8 shows the output query (**.sql**) file that results from the processing of this file by **+orndml**. An ORN header, which declares and initializes ORN Additive system objects, has been included at the beginning of the query file. Also, ORN Additive DML statements have been replaced with corresponding T-SQL DML statements and statements that execute ORN Additive system stored procedures.

```
-- Begin +<> (ORN Additive code)
  -- ORN Header (generated from +ornddl, version 1.05)
     --- ------ -------------------------------------
  -- Defines session temporary tables needed for changes to ORN-related tables.

  -- This code must be included at the beginning of a query at the outermost level
  -- and executed before any INSERT, DELETE, or UPDATE statements are executed on
  -- ORN-related tables and before any +<> statements are executed.  If this code
  -- is generated by +ornddl, it supports changes to only a single database.  If
  -- generated by +orndml, it supports changes to all databases referenced in the
  -- +<> USE statements given within the preprocessed .sql+ file.
  ...
-- End +<> (ORN Additive Code)

-- Begin +<> (+<> code generated by +orndml, version 1.05.)
  USE CompanyDB;
-- End +<>
GO

-- Begin +<>
  EXEC @ORN_Error = master.dbo.usp_ORN_PreBeginTran;
  IF @ORN_Error = 0
    BEGIN
    BEGIN TRAN;
    EXEC master.dbo.usp_ORN_PostBeginTran '';
    END;
-- End +<>
  INSERT INTO Carpool VALUES ('Airport', ...);
  UPDATE Employee
    SET carpoolId = 'Airport' WHERE ssn = '444-44-4444' OR ssn = '777-77-7777';
    -- Will implicitly delete any carpool if reassignment of an employee to the 'Airport' carpool
    --    results in less than two employees.
BEGIN TRY
 -- Begin +<>
   EXEC @ORN_Error = master.dbo.usp_ORN_PreCommitTran;
   IF @ORN_Error = 0
     BEGIN
     COMMIT TRAN;
     EXEC master.dbo.usp_ORN_PostCommitTran;
     END;
 -- End +<>
    -- Will raise exception if two employees have not been assigned to the 'Airport' carpool.
   PRINT 'Transaction committed.';
END TRY
BEGIN CATCH
 -- Begin +<>
   EXEC @ORN_Error = master.dbo.usp_ORN_PreRollbackTran '';
   IF @ORN_Error = 0
     BEGIN
     ROLLBACK TRAN;
     EXEC master.dbo.usp_ORN_PostRollbackTran '';
     END;
 -- End +<>
   Print 'Database transaction rolled back';
END CATCH
GO
```

Fig. 6.8 Query (.sql) file resulting from +orndml processing of query (.sql+) file shown in Fig. 6.7

6.4 Conclusion

Adding ORN to a relational database, which ORN Additive demonstrates is feasible, can increase the productivity of developing database systems and provide improved data integrity in these systems. Database implementation is made easier as database models map more directly to SQL, facilitating a more model-driven development approach (Mellor et al. 2003). More specifically, ORN Additive shows that:

- The multiplicities and bindings for an association in an ORN-extended UML class diagram can be mapped directly into an *<association>* specified for the foreign key that implements the association in SQL. This was indicated by the example database model and corresponding T-SQL/ORN Additive database definition given in this chapter and becomes even more apparent in Chapter 8.
- Association semantics more powerful than those found in standard SQL can be easily declared in SQL and enforced by the DBMS. Such semantics, for example, enforce multiplicities, delete a row in a table when a given number of rows in the referencing table no longer reference it, or implicitly delete a row in a table when it is no longer used, i.e., when there are no longer any required references to it.
- Database implementations can be made easier and less error-prone in that there is less need for database developers to code, test, and maintain complex constraint and trigger specifications.

Obviously, the enforcement of ORN semantics by the T-SQL code generated by ORN Additive requires overhead and thus impacts the performance of the T-SQL statements INSERT, DELETE, UPDATE, and COMMIT TRAN. Much of this overhead exists because ORN Additive implements ORN via triggers "on top of" the DBMS, here SQL Server. **Ideally, there would no need for an "ORN Additive"!** ORN implementation would be integrated within SQL and the DBMS resulting in a much more efficient implementation. Such an implementation is discussed in Chapter 11.

ORN Additive is fully functional and testing to date has revealed no existing problems in implementing ORN. It is continually being used for student database projects, but "real users" are needed to employ and test the tool on real-world databases. Any feedback will be much appreciated.

Chapter 7

Object Relater *Plus* (OR+)

An ORN-Extended Object DBMS

Object Relater Plus (OR+) adds ORN to ObjectStore (Progress 2006) but does so via two new languages, the Object Database Definition Language (ODDL) and the Object Database Manipulation Language (ODML). These languages extend and imitate an ODMG standard for object databases (Cattel et al. 2000, 1997). Adding ORN to ObjectStore increases the productivity of developing object database systems and improves data integrity in these systems. Implementation is easier as database models more directly map to the required ODDL statements of OR+. More specifically, relationships can be defined with the multiplicities and bindings that are defined in UML class diagrams. These multiplicities and bindings are automatically enforced by the database system as database changes are made using ODML. OR+ was developed to show the feasibility of defining associations to an ODBMS using ORN.

Although OR+ was developed as a prototype, it can be a very productive tool for developing many object database applications. Many lines of C++ code, involving complex class methods, are automatically generated by the tool. This code would otherwise have to be painstakingly implemented, tested, and maintained by database developers. The complete OR+ tool, including the *OR+ ODDL User's Guide* and the *OR+ ODML User's Guide* can be downloaded from the author's website at the URL given in the Preface. The user's guides provide the technical details for installing and using the tool.

This chapter provides an overview of OR+, first discussed in Ehlmann and Riccardi (1997b). It is partially based on this paper[1] and the OR+ user's guides. Section 7.1 examines OR+ capabilities and explains its level of compatibility with the ODMG standard. Sections 7.2 and 7.3 provide detailed overviews of ODDL and ODML, respectively. Section 7.4 discusses system architecture, implementation, and extensibility. Section 7.5 concludes by summarizing the benefits and limitations of OR+, giving its status, and indicating how its limitations can be easily overcome.

7.1 Capabilities and Compatibilities

The capabilities of OR+ are provided by the ODDL processor and the ODML application program interface (API), which is implemented as a layer on top of Object-

[1] Portions reprinted, with permission, from "Object Relater *Plus*: a practical tool for developing enhanced object databases," *Proc Data Eng Conf*, IEEE Computer Society, 412-421.

B.K. Ehlmann, *Object Relationship Notation (ORN) for Database Applications*,
Advances in Database Systems 39, DOI 10.1007/978-0-387-09554-7_7,

Store. This section briefly describes how ODDL and ODML essentially add ORN to ObjectStore and extend the ODMG standard with some additional capabilities.

ODDL was developed as a compatible extension of the ODMG 2.0 C++ Object Data Language (ODL), though its relationship declarations reflect the higher-level, language-independent ODMG 2.0 ODL. ODDL, like ODL, is an object DDL because it allows object behavior to be specified as C++ class member functions (or methods). ODML is a compatible extension of ODMG 2.0 C++ Object Manipulation Language (OML) (Future Binding, Release 1.1) and provides database manipulation based on an ODDL-defined database. ODDL and ODML provide a "facade" for database development that is ODBMS vendor-independent. ObjectStore is the structure beneath this facade, supplying much of the basic ODBMS capability.

ODDL primarily describes an object database at a conceptual level in an ANSI-SPARC sense (ANSI 1975). Footnotes, however, allow implementation details, i.e., an internal schema, to be given. Sidebars facilitate the specification of an external schema and extended capabilities for a specific application domain, e.g., scientific databases. The separation of internal, conceptual, and external schema and ODBMS extensibility via sidebars are not capabilities found in the ODMG standard.

The most important capability, however, that OR+ adds to the ODMG standard (and ObjectStore) is ORN. ODDL allows an *<association>* to be given for a relationship, represented as an object-based attribute pair, e.g., carpool and riders in Fig. 7.1. The ODML API extends ODMG OML to support ORN. This means that ORN semantics are automatically maintained by the ODBMS, which in turn means that ODML operations are not primitive object operations but are complex object operations. Fig. 7.2 shows a C++ program segment that uses ODML types Database, Transaction, and d_Iterator and operations open, begin, lookup_object, next, Delete, commit, and close to delete an employee (and possibly a carpool) from the CompanyDb database.

```
Database CompanyDb { ...
  class Carpool;
  class Employee {
    d_String          ssn;    // Social Security Number
    ...
    Carpool           carpool inverse riders ?<2..15-to-0..1>;
$C++
    void RaiseSalary(int percentage);
$.
  };
  extent Set<Employee> Employees;
...
  class Carpool {
    d_String          id;
    ...
    Set<Employee>  riders inverse carpool;
  };
  extent Set<Carpool>  Carpools;
  ...
};
```

Fig. 7.1 Example of a partial ODDL specification for a company database

```
...
Database CompanyDb;
Transaction t;
CompanyDb.open("co_db");
t.begin();
Employees_Ptr = (d_Set<Employee*>*)CompanyDb.lookup_object("Employees");
Employee* pE;
d_Iterator<Employee*> iE(*Employees_Ptr);
while (iE.next(pE))
  if (pE->ssn == "555-55-5555") {
    pE->Delete(); // Results in deletion of a Carpool object if lowerbound 2 is violated.
    break;
  };
t.commit();
CompanyDb.close();
...
```

Fig. 7.2 Example of a C++ code using the ODML API to update database CompanyDb

7.2 ODDL

The next two pages include figures referenced in the remainder of this chapter. Fig. 7.3 shows an example of a database model for a company database. The partial ODDL specification that was given in Fig. 7.1 was developed based on this model, and an expanded version of this specification is given in Fig. 7.4. Fig. 7.5 shows an example of a database model that is part of a nuclear physics experiments database. The partial ODDL specification in Fig. 7.6 is based on this database model.

7.2.1 ODDL specification

An ODDL specification consists of a number of forward class declarations, class definitions, and extent definitions embedded in a database definition. Control commands interspersed within the specification provide sidebars, footnotes, and related C++ code. Although ODDL is like ODMG ODL in syntax and semantics, a different terminology is used here to describe it.

A class definition defines a type of object, e.g., class Employee in Fig. 7.4. Objects have an identity independent of their attribute values. The isa clause specifies an "is a" relationship between a subclass and one or more superclasses, e.g., class Root isa Track in Fig. 7.6. C++ multiple inheritance semantics are operative for the "is a" relationship. A class definition usually defines one or more attributes, which are discussed later in this section.

An *extent* is a collection that allows all objects of a class to be directly accessible, often providing an entry point into the database, e.g., extent Set<Employee> Employees in Fig. 7.4. Such direct access to class objects is often desired for classes representing strong entity types.

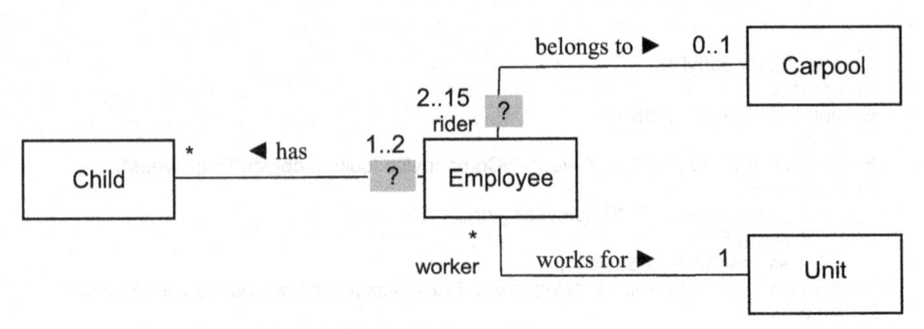

Fig. 7.3 Model for a company database

```
Database CompanyDb
$C++
 class Date{ ... };
$.
{
 class Unit;
 class Child;
 class Carpool;

 class Employee {
  d_String     ssn;  // Soc. Sec. No.
  d_String     name; // Last, First name
  int          type;  // 0 – perm, 1 – temp
  ...
  Unit         unit inverse workers <*-to-1>;
    // The introductory ODMG ODL key-
    // word "relationship" is optional.
  List<Child>  children inverse parents
               ?<1..2-to-*>;
  Carpool      carpool inverse riders
               ?<2..15-to-0..1>;
$C++
  void RaiseSalary(int percentage);
  void Print();  // Print employee.
  ...
$.
 };
 extent Set<Employee> Employees;
```

```
 class Unit {
  d_String         name;
  d_String         location; // building no.
  Set<Employee>    workers inverse unit;
                   // <1-to-*> by default
$C++
  void Print();  // Print unit and its employees.
  ...
$.
 };
 extent Set<Unit>  Units;

 class Child {
  d_String         name;
  Date             birthday;
  Set<Employee>    parents inverse children
                   <*-to-1..2>?;
 };

 class Carpool {
  d_String         id;
  ...
  Set<Employee>    riders inverse carpool;
 };
 extent Set<Carpool>  Carpools;
 ...
};
```

Fig. 7.4. Partial ODDL specification for the database model in Fig. 7.3

Comments can be given in ODDL at specific places for each *metaobject*—the database, a class, an extent, or an attribute—e.g., in Fig. 7.4, // Soc. Sec. No. for attribute ssn in class Employee. Unlike comments in ODMG ODL, these comments become part of the ODDL metadatabase and provide valuable information to a user who wishes to browse the metadatabase to discover database contents.

References can be attached to metaobjects. They reference footnotes, e.g., [1] for attribute subtracks within class Track in Fig. 7.6, or they reference sidebars, e.g., |1 for numberOfOffshoots within class Root. Footnotes and sidebars are discussed in Section 7.2.3.

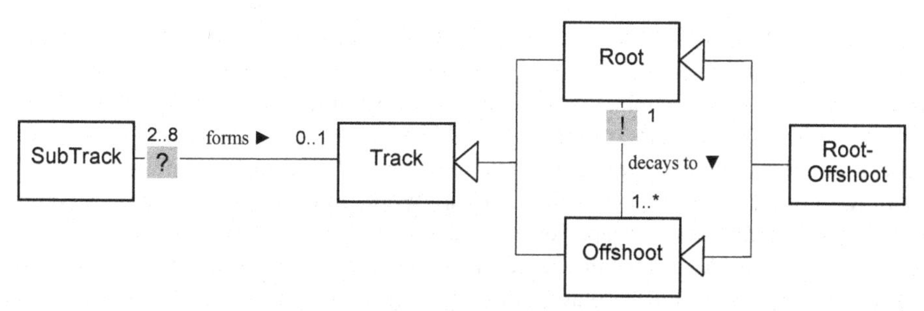

Fig. 7.5 Model for part of a nuclear physics experiments database

```
Database Experiments // for nuclear physics
{
$sidebar 1
derived
$.
 ...
 class Track;
 class Root;
 class Offshoot;
 class Subtrack { // particle subtrack in a
                  //  detector device
   Device device;
   Track  track inverse subTracks
          ?<2..8-to-0..1>;
   ...
 };
 class Track { // particle track within a
               //  sector of the detector
   Sector             sector;
   Collection<subtrack> subTracks
                        inverse track; [1]
   ...
 };
 class Root isa Track { //or :public track
   Collection<Offshoot> Offshoots
                        inverse Root
                        !<1-to-1..*>; [2]
   int numberOfOffshoots; |1
   ...
 };
 class Offshoot isa Track {  // track from a
                             //  decaying particle
   Root  root inverse Offshoots;
   ...
 };
 class RootOffshoot isa Root, Offshoot {
   ...
 };
...
$footnotes
[1] List
[2] Set
$.
};
```

Fig. 7.6 Partial ODDL specification for the database model in Fig. 7.5

Attributes serve to describe objects and within a class specify attributes common to all objects of the class. Attributes are *value-based* or *object-based.*

A value-based attribute, e.g., name within class Employee in Fig. 7.4, stores values that have no value-independent identity in the database. The attribute can be *single-valued* or *multi-valued.* Its type can be a C++ primitive type (e.g., int), a type defined by ObjectStore or OR+ (e.g., d_String), or a type previously defined in the ODDL specification via C++ code (e.g., an enum, a typedef, or a class like Date in Fig. 7.4).

An object-based attribute, e.g., unit within class Employee in Fig. 7.4, may reference one or more objects in the database. A pair of object-based attributes that are inverses of each other represent an association in ODDL. Each attribute represents one direction of a *bi-directional association.* (In the ODMG standard these object-based attributes are called relationship *traversal paths* and are preceded by the keyword relationship. In ODDL this keyword is optional.) The object-based attribute in a *subject class* references objects of a *related class* and represents the association from the viewpoint of a subject class object. For example, the attribute unit in class Employee, the subject class, references an object in class Unit, the related class, and represents the association between employees and units from the viewpoint of an Employee object. Its inverse attribute, Employees in class Unit, references zero, one, or more employees and represents this association from the viewpoint of a Unit object. An object-based attribute name in a subject class should describe the role a related class object or objects play in describing a subject class object—e.g., one or two Employee objects play the role of parents in describing a Child object.

When the association multiplicity (see Section 7.2.2 below) for the related class allows at most one related object, the type of an object-based attribute is a *type* reference where *type* is the name of the related class. The attribute type is a collection of such references when the multiplicity allows more than one related object. An object-based attribute type could always be defined as a collection, but the collection would not be fully utilized for a multiplicity upper bound of one unless the upper bound was later increased. The type of a collection may be a Set, which is unordered, or a List, which is ordered. Neither type can contain duplicate references. Unlike the ODMG standard, a collection type can also be given generically as a Collection. This type may then either be made more specific via a footnote, as in Fig. 7.6 for subTracks and Offshoots, or be allowed to default to a Set. An object-based attribute that is a reference is of course *single-valued*, one that is a collection is *multi-valued.*

7.2.2 <association>

An *<association>*, e.g., <*-to-1> for unit in class Employee in Fig. 7.4, describes the semantics of an association as represented by an object-based attribute and its inverse attribute. The multiplicities and bindings of an *<association>* can be mapped directly from an ORN-extended class diagram. The syntax of an *<association>* is, of course, that defined in Fig. 2.1. Table 7.1 defines the semantics in the context of the object model as defined by ODDL. The bindings and multiplicities in an *<association>* define the semantics of the association, which in turn determine the scope of complex and composite objects.

An *<association>* may be given for one, both, or neither attribute of an inverse attribute pair. If given for just one, the inverse of the given *<association>* is implicitly assumed for the other—e.g., in Fig. 7.4, <0..1-to-2..15>? is assumed for riders in class Carpool. If given for both attributes, each *<association>* must be the inverse of the other—e.g., in Fig. 7.4, <*-to-1..2>? and ?<1..2-to-*> for the parents and children attributes, respectively, are inverses of each other. If not given for either attribute, a default *<association>* is assigned to each based on whether the attributes are single-valued or multi-valued. If single-valued, the default *<association>* is <...-to-0..1>. If multi-valued, it is <...-to-*>.

Table 7.1 ORN semantics for the object model of ODDL

Semantics are given in terms of a subject class *S* with multiplicity *m* and binding *b* in an association *A* with a related class *R* (which could be *S* in a different role).

A is implemented by an object-based attribute *roleR* and inverse object-based attribute *roleS*. The type of *roleR* is *R* or some suitable collection of *R* (shown here as *R*), and the type of *roleS* is *S* or some suitable collection of *S* (shown here as Set<*S*>). The multiplicity for *S* (or *roleR)* is *m*, and the binding is *b*. An *A* link, now an "*A* reference pair," is destroyed when a *roleS* or *roleR* reference is destroyed or changed to reference a new object. Creating or destroying either reference is equivalent to and includes creating and destroying, respectively, its inverse reference.

```
class S {
   ...
   R roleR inverse roleS b<m-to-...;
}

class R {
   ...
   Set<S> roleS inverse roleR;
}
```

***<multiplicity>*:** Semantics are the same as those in UML. Essentially, *m* indicates a lower bound and upper bound on the number of objects of type *S* that can be related via *A* to each object of type *R*.

An *R* object can be created provided this does not violate *m*. The check for a lower bound violation is deferred until transaction commit. A *roleR* reference can be created provided this does not violate *m*. The check for an upper bound violation is immediate. The enforcement of *m* on the deletion of an *S* object or destruction of a *roleR* (or *roleS*) reference is determined by the binding *b*.

***<binding>*:** A **|** in *b* denotes a "cut" and an Implicit, i.e., system initiated, destruction of an existing *A* reference that must occur on deletion of an *S* object. An **X** in *b* denotes a "cross out" and an eXplicit, user initiated, destruction of an *A* reference.[1]

An *S* object deletion and an explicit *A* reference destruction are *complex object operations*. Deletion of an *S* object succeeds only if all existing association references involving that object are implicitly destructible. Also, deletion of an *S* object or destruction of an *A* reference succeeds only if all required implicit object deletions succeed.

***<di>*:** A destructibility indicator in *b* specifies the destructibility of an *A* (i.e., *roleR* and inverse *roleS*) reference. The meaning of each indicator is given below. This meaning can alternatively be described by the actions taken on an attempt to destroy an *A* reference. These actions are given in brackets. If a *<di>* is given after a **|**, it applies to implicit reference destruction; if given after an **X**, it applies to explicit reference destruction; and if given alone, it applies to both. If a *<di>* is not given, i.e., is nil, for implicit reference destruction, explicit reference destruction, or both, default destructibility applies to whichever.

nil — *Default destructibility*. A reference can be destroyed provided this does not violate *m*.[2] [Destroy the reference. If *m* is violated[2], raise an exception[3].]

\- — *Negative destructibility*. A reference cannot be destroyed. [Raise an exception.[3]]

~? or **?** — *Conditional cascade destructibility*. A reference can be destroyed, but if this violates *m* (**?**), the destruction must be cascaded (~) to the related *R* object, i.e., this object must be implicitly deleted. [Destroy the reference. Delete the related *R* object**?** If *m* **is violated**, yes; else no.]

~! or **!** — *Emphatic cascade destructibility*. A reference can be destroyed, but the destruction must be cascaded (~) to the related *R* object. [Destroy the reference. Delete the related *R* object**!**]

~' or **'** — *Tentative (or qualified) cascade destructibility*. A reference can be destroyed, but an attempt must be made to cascade (~) the destruction to the related *R* object; however, this implicit *R* object deletion must be undone if it fails, but is required if and only if its undoing would violate *m*.[2] (Think of the **'** as a "pruned back !" or as a "qualifying footnote reference" on the cascade.) [Drop the reference. Delete the related *R* object.**'** (**'** – If an exception occurs, undo the delete and then, if *m* is violated[2], raise an exception[3].)]

---

1 - A *roleS* reference change done as a single operation that replaces an *S* object with another is not treated as an explicit link destruction relative to class *S* (but is relative to *R*) and is allowed if allowed by other multiplicities and bindings.

2 - The check for a lower bound violation is deferred until the end of the current complex object operation.

3 - The current complex object operation is undone.

The representation of associations as object-based attributes in ODDL, and in object databases in general, permits the user to conveniently view, query, and manipulate an association from the context of a specific object. In ODML, association creation, destruction, and change occur via operations on the object-based attributes. An association instance is created when the value of a single-valued attribute that is null is set to reference an object or when an object reference is inserted into a multi-valued attribute. An association instance is destroyed when a non-null value of a single-valued attribute is set to null or when an object reference is removed from a multi-valued attribute. An association instance is changed when the non-null value of a single-valued attribute is set to reference a different object or when an object reference in a multi-valued attribute is replaced by another. Inverse attributes and all association semantics as described by ORN are automatically maintained as object deletions occur and as association creations, destructions, and changes are made via object-based attributes. These operations are illustrated in more detail in Section 7.3.

7.2.3 Control commands

Control commands in ODDL provide a number of auxiliary functions. They permit the declaration of attribute types and object behavior to be given directly in C++. They also provide sidebars and footnotes and permit the selective processing of an ODDL specification.

The $C++ and $. commands delimit C++ code. Depending on placement within the ODDL specification, this code can provide enum, typedef, struct, and class definitions that are local or global to a class, e.g., class Date in Fig. 7.4. The code can also provide private and public class data member and function member declarations, e.g., RaiseSalary() for class Employee in Fig. 7.4. ODDL places code delimited by $C++ and $. into a metadatabase code library and appropriately generates it into .h file(s), which must be included in program code that accesses the database.

The $sidebar and $. commands delimit a sidebar. Sidebars provide high-level application related specifications that extend the conceptual description of the object database provided by ODDL. These specifications may define external views or give options, e.g., derived in Fig. 7.6, that are relevant to a specific or generic application, e.g., scientific databases. Sidebars are given in a language designed by the OR+ user and parsed using an ODDL supplied scanner prior to ODDL code generation (see Section 7.4.3). Once a sidebar is defined, it can be referenced via its number by subsequent ODDL metaobjects.

The $footnote(s) command indicates that one or more footnotes follow on subsequent lines. A $. terminates the footnote(s). Footnotes provide low-level implementation details, i.e., an internal schema. These details may specify the object database storage structures for collections, e.g., [1] List in Fig. 7.6, or provide for random access of collections via keys and sequential access via prescribed orderings. The integer identifying a footnote must, of course, match one or more previous footnote references.

7.3 ODML

ODML provides a C++ API for an object database defined by ODDL. It is a collection of C++ macros, classes, and functions. Fig. 7.7 shows a sample program that uses ODML to access and manipulate the database defined by the ODDL specification given in Fig. 7.4. The discussion below focuses on this sample program.

The program adds a public relations unit to the company database and a new employee, who is to work for this new unit. In addition, it deletes all temporary employees and reassigns all employees in the customer support unit to the public relations unit. It then prints all units, including their employees. The following paragraphs explain and comment on segments of the program.

```
 1 // sample_prog.C
 2
 3 #include <iostream.h>
 4 #include "CompanyDbOR+.h"
 5
 6 void PrintUnits() // Print all units, including the employees working for each unit.
 7 {
 8   cout << "Company Employees by Unit:" << endl;
 9   d_Iterator<Unit*> iU(*Units_Ptr);
10   unit* pU;
11   while (iU.next(dp))  pU->Print();
12 }
13
14 void main()
15 {
16   Database db;
17   Transaction t;
18   db.open("co_db");
19   t.begin();
20   Units_Ptr = (d_Set<Unit*>*)db.lookup_object("Units");
21   Employees_Ptr = (d_Set<Employee*>*)db.lookup_object("Employees");
22
23   Unit* pU = NEW(db, Unit)("Public Relations");
24   pU->location = "205 Building B";
25   Employee* pE = NEW(db, Employee)("Doe, John Q.");
26   pE->unit = pU;  // or pU->Employees.insert_element(pE);
27   ... // Set additional employee attributes.
28
29   d_Iterator<Employee*> iE(*Employees_Ptr);
30   while (iE.next(pE))
31      if (pE-> type == 2)
32        pE->Delete() // May result in the deletion of carpool and child objects.
33      else
34        if (pE->unit->name == "Customer Support")  pE->unit = pU;
35
36   PrintUnits();
37
38   t.commit();
39   db.close();
40 }
```

Fig. 7.7 Sample program that uses ODML to modify the database defined by the ODDL in Fig. 7.4

The Units_Ptr and Employees_Ptr (lines 20 and 21) are declared in CompanyDbOR+.h, which is generated by ODDL. These pointer variables are set to reference the extents Units and Employees, respectively, which are *named objects* in the database. Named objects provide entry points into the database, i.e., objects through which other persistent objects can be accessed.

The NEW macro (line 23) calls an ODML overloaded new operator that creates a persistent object of the given type in the given database and calls a constructor. The constructors for classes Employee and Unit are not shown in Fig. 7.4.

Assigning John Q. Doe to public relations (line 26), i.e., creating a relationship between the new Employee object and the public relations Unit object, can be done by either setting the attribute unit for the new Employee object to reference the public relations object or by inserting a reference to the new Employee object into the attribute employees for the public relations object. To reflect changes made to an object-based attribute, OR+ automatically makes corresponding changes to the inverse attribute. The choice of perspective for making changes is the programmer's.

Iterators (line 29) are provided by ODML to sequentially access the elements of a collection. In the sample program, the d_Iterator named iE provides such access to all objects of the Employees extent.

Deletion of a temporary employee via pE->Delete() (line 32) causes more than just the deletion of the employee object. All relationship semantics associated with an employee, i.e., those modeled in Fig. 7.3 and defined in Fig. 7.4, are enforced. This means that relationship references are implicitly destroyed—perhaps, for example, the employee object's reference to a carpool—and other objects are implicitly deleted—perhaps one or more child objects and/or a carpool object.

The assignment pE->unit = pU (line 34) performs a relationship change from the perspective of an employee, i.e., it transfers an employee from customer support to public relations. Again, the inverse attribute employees for both the customer support and public relation objects is automatically updated.

The function PrintUnits() (lines 36 and 6) iterates through the Units extent calling the Print() method of class Unit on each Unit object (line 11). This method (declared in Fig. 7.4) iterates through the employees attribute, invoking the Print() method of class Employee on each employee referenced by the attribute. The definitions of these methods must of course be linked to the sample program.

The call to t.commit() (line 38) commits the changes made by the program. The commit() method ensures that all lower bound multiplicities are met for objects created during the transaction, e.g., that unit for the John Q. Doe object is not 0.

With ORN it is sometimes necessary to temporarily suspend lower bound multiplicity checks to allow relationships among objects to be reshuffled. For example, assume in Fig. 7.3 a new |~<1-to-1>, "is assigned" association between the class Employee and a new class OfficeKey. This new class defines objects that correspond to a physical key and also record information about the assignment of an office to an employee. In the ODDL of Fig. 7.4, an object-based attribute officeKey is added to the Employee class to reference an OfficeKey object. Now, suppose that a decision is made to reshuffle the offices assigned to three employees. Employee (referenced by) e1 currently has office key (referenced by) k1, e2 has k2, and e3 has k3. k1 must be reassigned to e2, k2 to e3, and k3 to e1. This can be done by the code in Fig. 7.8 placed within a Transaction t. The function SetRXCmode() sets the Relationship eXChange (RXC) transaction mode. ResetRXCmode() resets this mode. When RXC mode is set, lower bound multiplicity checks are not performed as usual on completion of complex relationship destruction (e.g., e1.officeKey = 0;) and change operations and binding actions that might result from such checks are not taken. Instead, relevant attributes, e.g., officeKey above, are checked on commit of the application-specified transaction to ensure that lower bound multiplicity constraints are satisfied. Exceptions result if they are not. Without the RXC mode, certain manipulations on some associations types could not be done without violating multiplicities.

```
t.SetRXCmode();
e1.officeKey = 0;
e2.officeKey = 0;
e3.officeKey = 0;
e1.officeKey = &k3;
e2.officeKey = &k1;
e3.officeKey = &k2;
t.ResetRXCmode();
```

Fig. 7.8 Using RXC mode to exchange relationships

7.4 Architecture, Implementation, and Extensibility

This section provides insight into OR+'s architecture, implementation, and how it can be extended by a user. OR+ was developed for two types of users—one who uses ODBMS for a specific database application and the other who uses it as a tool to develop an extended ODBMS, often adding capabilities for a particular application domain. OR+ has been compiled with multiple C++ compilers and run on multiple Unix platforms, although currently only one compiler and platform is supported, which is the Sun C++ compiler and the Sun Solaris platform. (Source code may be obtained on request, and the system can be ported to other Unix platforms.)

7.4.1 Architecture

Fig. 7.9 shows the architecture of the OR+ system. It indicates the processes and files needed to create an ORN-enhanced ObjectStore database and run a database application program that accesses and/or updates it. What is provided by OR+ is shown in bold. What must be developed by the OR+ user is shown in italics. What must be acquired by the OR+ user is ObjectStore.

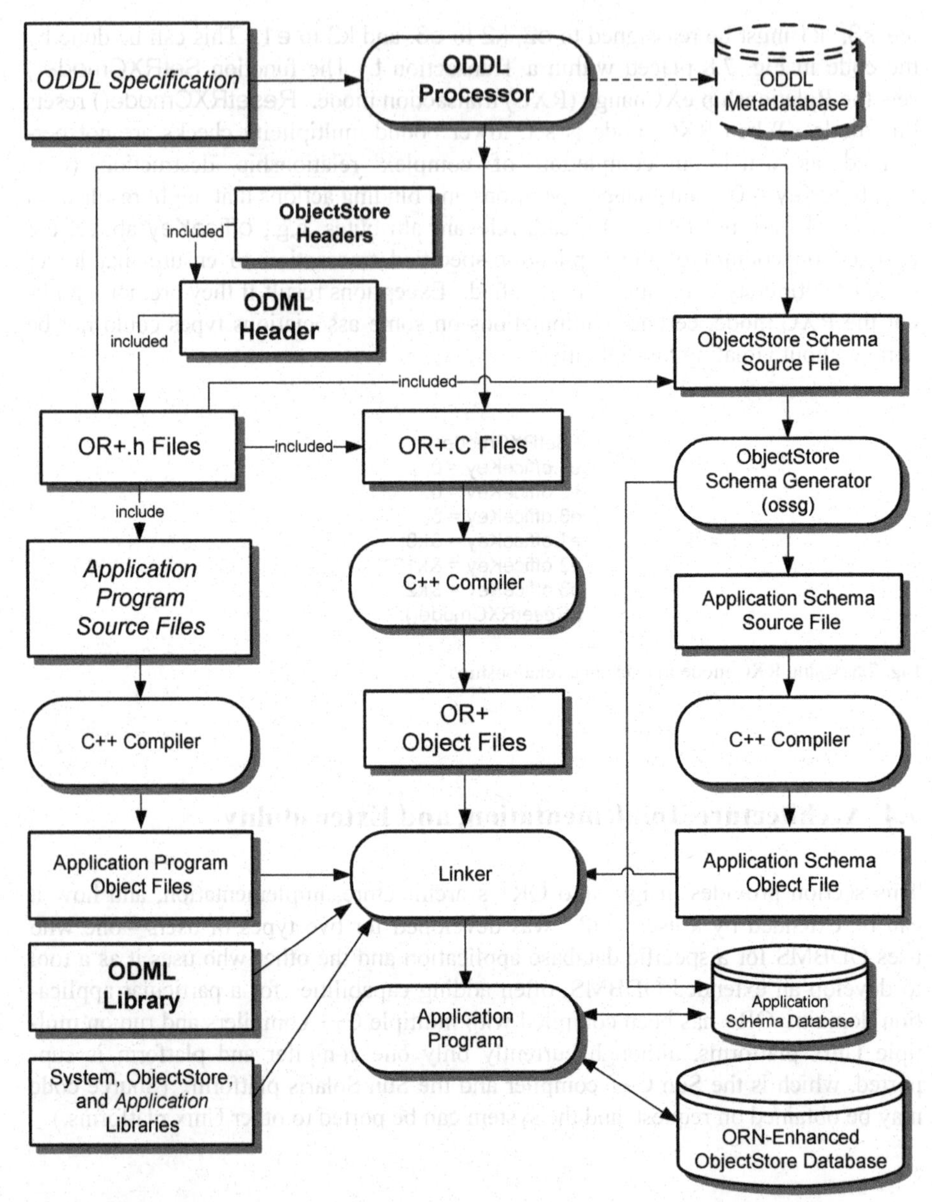

Fig. 7.9 OR+ architecture

The ODDL processor of OR+ is invoked by the oddl command, which includes a number of options. Typical processing verifies proper syntax and semantics of the ODDL specification, checks for inconsistencies in the *<association>*s given for relationships (see Section 9.1), and assigns default *<association>*s as required. The ODDL processor builds a metadatabase, which can optionally be saved as an ObjectStore database and later processed by ODDL, the ObjectStore browser, or programs and queries written by the OR+ user. An error-free specification normally results in the generation of a number of C++/ObjectStore source files. The database-specific OR+.h and OR+.C files implement ODML when included in application program source files that are compiled and linked with the database-independent ODML library and ObjectStore libraries. The ODDL-generated ObjectStore schema source file is (as prescribed by ObjectStore) processed by the ObjectStore Schema Generator (ossg) to create the ObjectStore application schema database and the application schema source file. This latter file is compiled to create the application schema object file, which must be linked with the application program. Makefiles supplied by OR+ and ObjectStore automate the required processing.

7.4.2 Implementation

As previously indicated, ODDL and ODML provide an ODMG-compatible interface to ObjectStore and enhanced ODBMS capabilities like ORN. To create this interface, ODDL classes and member functions "wrap" analogous ObjectStore classes and member functions.

The addition of ORN to the capabilities of the ODMG standard (and ObjectStore) add surprisingly little code complexity and overhead to that required to maintain inverse attributes, which is already done by most commercial ODBMSs. The approach used extends a fairly standard one to implementing relationships in an ODBMS and places little additional overhead on the <0..1-to-0..1>, <0..1-to-*>, and <*-to-*> associations, which are the ORN association types generally supported by existing ODBMSs. Within an object, each object-based attribute having an inverse clause is an instance of its own specific relationship class. Via operator overloading, the attribute is manipulated just as if it were a normal C++ pointer or collection (of pointers). Overloaded operators, like assignment on a pointer and remove_element() on a collection, implement ORN semantics as do an overloaded new operator and a Delete() operation defined for all ORN-related objects. These operations work in conjunction with begin(), commit(), and abort() operations on an ODML Transaction object (which contains a corresponding ObjectStore transaction object).

To gain some understanding of how relationship semantics are enforced, assume that pE->Delete() has just been executed in the program in Fig. 7.7 (line 32), i.e., an attempt has been made to delete an Employee object. The following is a simplified trace of OR+ processing.

First, the ODDL generated Delete() method for class Employee, the subject class, will for each object-based attribute in the class, e.g., carpool, check if any references exist and, if so, call a generated method on the relationship class for the attribute to remove them. That is, an attempt is made to implicitly destroy all relationship references involving the Employee object. The object is deleted if and only if no exceptions result.

The relationship class method for a specific object-based attribute (cited above) calls an ODML library method, inherited from a class representing a generic object-based attribute, passing to it the encoded *<association>* for the specific object-based attribute. For each existing reference in the attribute, this method removes it and calls an inherited library method on the relationship class for the inverse attribute, here riders, in the referenced (or related) object, here a Carpool object, to remove from the attribute the reference to the Employee object. It then calls yet another inherited method to check the multiplicity and implicit destructibility binding for the Employee class in the *<association>*. Actions are taken as described in Table 7.1. If, for instance, the binding is |-, an exception is raised. If, however, it is |?, as for Employee in the "belongs to" association, and the lower bound cardinality, here 2, is violated, then an implicit form of Delete() is called on the related object, here a Carpool object, and this whole process begins anew for this object.

Before addressing ODDL extensibility, a few additional details about the implementation of ORN are important to note. First, this implementation does not affect user-written constructors and destructors for ORN-related classes. Multiplicity checks and implicit relationship destructions and object deletions are not done within constructors and destructors (unless they result from explicit relationship destructions and object deletions done by the application within constructors). Second, association semantics are maintained by run-time interpretation of encoded *<association>* objects, which are instantiated as global variables within ODDL-generated header files included in the application program. Third, certain sets of object references are maintained by the system during the processing of a transaction. These sets are used to detect circularities that can result when implicit object deletions are propagated. They are also used to keep tabs on all newly created objects and reference destructions requiring deferred lower bound multiplicity checks so that these checks can be done on transaction commit. Finally, the actual removal of references from collections and the deletion of primitive objects are done by ObjectStore.

More information on the implementation of ORN in OR+ can be gleaned from the algorithms given in Chapters 10 and 12.

7.4.3 Extensibility

An option on the oddl command and a number of ODDL supplied "tools" facilitate the extension of OR+ into an even more enhanced ODBMS. When the -g option is specified, user-written code in OR+ assumes control of ODDL code generation. The tools supplied to the user for extending ODDL are a scanner, hooks in the metadatabase, a code generation facility, public member functions and iterators that facilitate access to metadatabase objects, and a class representing the ODDL provided database system layer.

As previously indicated in Section 7.2.3, the scanner is provided for parsing sidebars. The sidebars are initially stored in the ODDL metadatabase as text. The first step in generating a user-enhanced object database system is to parse these sidebars, appropriately issuing errors, if any, and updating an extended ODDL metadatabase.

Metadatabase extension is facilitated by hooks, i.e., pointers to metaobject extensions, that are included in each metaobject.

A C++ preprocessor, c^pp, is provided to facilitate code generation. This tool is illustrated by the code shown in Fig. 7.10, which generates for each extent in the ODDL metadatabase an external pointer definition and pointer declaration. When processed by c^pp, ^ (insert) commands are translated into << operations on a text file that is specified via the SetCodeFile() function. For the extent Employees, defined in Fig. 7.4, the code in Fig. 7.10 generates the following pointer declaration in an OR+.h file identified by *Hfile:

```
extern d_Set<Employee*>* Employees_Ptr;
```

It generates the following corresponding definition in a OR+.C file identified by *Cfile:

```
d_Set<Employee*>*     Employees_Ptr;
```

This is the Employees_Ptr that is used in the program given in Fig. 7.7.

```
list_cursor cE(Mdb->Extents); // Mdb points to the metadatabase
extent* pE;
while (cE.More()) {
  pE = (extent*)cE.Next();
  SetCodeFile(*Hfile);
^extern d_`pE->ICT`<`pE->ExtendedClass->Name`*>* `pE->Name`_Ptr;/
  SetCodeFile(*Cfile);
^  d_`pE->ICT`<`pE->ExtendedClass->Name`*>*    `pE->Name`_Ptr;/
}
```

Fig. 7.10 Using the c^pp preprocessor for code generation

The public member functions of an ODDL defined class, dbs_layer, enable the OR+ user to generate all portions of the ODDL-provided database system layer. This is the layer of software that the user is extending and essentially includes ODMG capabilities enhanced with ORN. The member functions of dbs_layer give the user the flexibility to generate portions of the database system layer into user-specified source files and to mix in user generated code at strategic points.

7.5 Conclusion

This chapter has described OR+ and emphasized features that make it a useful tool for developing enhanced object databases and extending ODBMSs with additional enhancements. The features of OR+ are summarized as follows:

- the facilitation of an object database development methodology where a database model represented by a UML class diagram is more directly mapped into an object DDL, e.g., ODDL (The precise rules for this mapping are given in Section 8.2.)

- a means, ORN, to better describe associations during analysis and design and define them to an ODBMS so that their semantics can be automatically maintained
- compatibility with a standard for ODBMSs and thus a degree of vendor-independence
- separation of external, conceptual, and internal ODB schemata via footnotes and sidebars
- extensibility via sidebars and supporting tools that facilitate generic and domain-specific enhancements

All of the OR+ capabilities described in this chapter have been implemented except for the following two:

- the derived sidebar, which was given in Fig. 7.6 as an example of a user-defined ODDL extension
- footnotes that specify keys, which were mentioned in Section 7.2.3

Implementing these capabilities would not be difficult.

They were not implemented, however, because OR+ is a prototype, originally developed to experiment with enhanced ODBMS features. Implementation is constrained by the user capabilities and architecture of the commercial ODBMS upon which it is built (which can be changed). In addition, some ODBMS capabilities are only available at the level of its underlying ODBMS, e.g., concurrency control and database schema migration. Obviously, overall implementation would be improved if all OR+ features were well integrated into a non-prototype ODBMS.

OR+ was first employed at the Supercomputer Computations Research Institute (SCRI) at Florida State University in the early 1990's to research enhanced object database capabilities and methodologies for scientific databases. Specifically, it was used to design a nuclear physics experiments database (Riccardi and Ehlmann 1991, Ehlmann 1992, Ehlmann et al. 1992, Ehlmann et al. 1993). Since then, it has been used for student database projects in graduate-level, DBMS courses. OR+ is the ODBMS that was used to implement the ORN Simulator (Chapter 3); therefore, its implementation of ORN is well tested.

Object database users, researchers, and vendors are encouraged to download OR+, experiment with it, and employ it to develop object database systems or ODBMSs. Any feedback will be much appreciated.

Chapter 8

Mapping Database Models to DDLs

From ORN-Extended Class Diagrams to ORN-Extended DBMSs

Chapter 6 described ORN Additive, which extends the SQL Server RDBMS with ORN. Chapter 7 described OR+, which extends the Object Store ODBMS with ORN. Both chapters presented an example of an ORN-extended class diagram that modeled a company database. Chapters 6 and 7 also gave DDLs that implemented the model as an ORN-extended RDBMS and ORN-extended ODBMS, respectively.

This chapter shows precisely how such DDLs are obtained, or mapped, from a database model. Section 8.1 shows how classes and associations in a database model along with ORN-defined association semantics are easily mapped to an ORN-extended SQL. Section 8.2 shows how such classes, associations, and semantics are easily mapped to an ORN-extended DDL for an ODBMS. Section 8.3 concludes with a brief summary remark.

8.1 Mapping an ORN-Extended Model to an ORN-Extended SQL

This section provides directions for mapping a database model given by an ORN-extended class diagram to an ORN-extended SQL. Here, the examples given use T-SQL with ORN Additive DDL statements. Chapter 11 describes another ORN-extended SQL, based on an SQL standard (ANSI 2008), for which the directions below also apply; the T-SQL given in examples can be easily translated.

The basic steps in the mapping process, which are given in greater detail in the subsections that follow, are:

1. Transform the database model, if necessary, into a model more compatible with the relational model by using appropriate association patterns.
2. Map classes to tables declaring appropriate attributes and primary keys.
3. Map associations to foreign keys declaring for each the same ORN as defined for the modeled association.

8.1.1 Transforming the model for a relational database

To simplify the mapping of the class diagram model to SQL, we first transform the model to make it more compatible to the way attributes and relationships must be implemented in a relational database.

B.K. Ehlmann, *Object Relationship Notation (ORN) for Database Applications*,
Advances in Database Systems 39, DOI 10.1007/978-0-387-09554-7_8,

8.1.1.1 Many-to-many and *n*-ary associations

Using the "is associated by" association pattern, given in Section 4.2.4, transform each many-to-many binary and each *n*-ary association where $n \geq 3$ into appropriate *n* one-to-many associations. Each one-to-many association is between one of the related classes and an associating class. This class is similar to and replaces the association class if one is given in the model for the original association. If not, create a new associating class, appropriately naming it to represent the original association. Each object of this associating class represents an instance of the original association and is associated with exactly one object from each of the related classes. Ensure that this set of related objects is unique for the associating class (as it was for the original association) by adding an appropriate constraint to the class diagram. Place any attributes for the original association into the associating class.

For example, transform the many-to-many association in Fig. 8.1 into an associating class Assignment; two one-to-many, "is associated by" association types; and an appropriate constraint as given in Fig. 4.6 (b). Section 4.3 gives an example of the transformation of another many-to-many association, between songs and CDs, into an associating class, two "is associated by" associations, and a constraint. The multiplicities and bindings of the original association are mapped to the new associations as shown in Fig. 4.17.

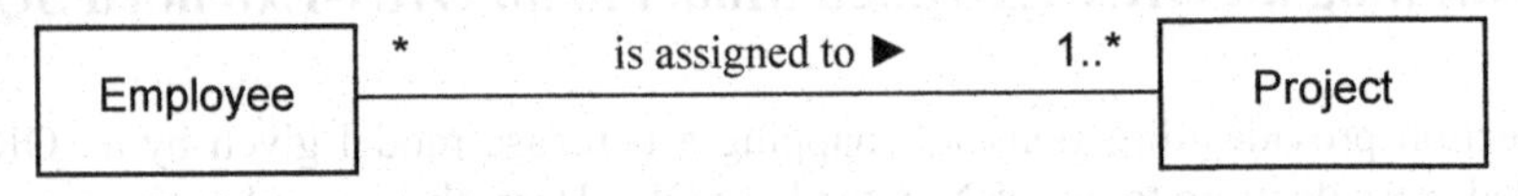

Fig. 8.1 Many-to-many association transformed into two one-to-many associations in Fig. 4.6 (b)

Fig. 8.2 shows the transformation of yet another many-to-many association, an intra-class association. The attribute relation indicates mother, stepmother, father, etc.. The ' and |- bindings included in the original association again show how bindings are transformed along with multiplicities.

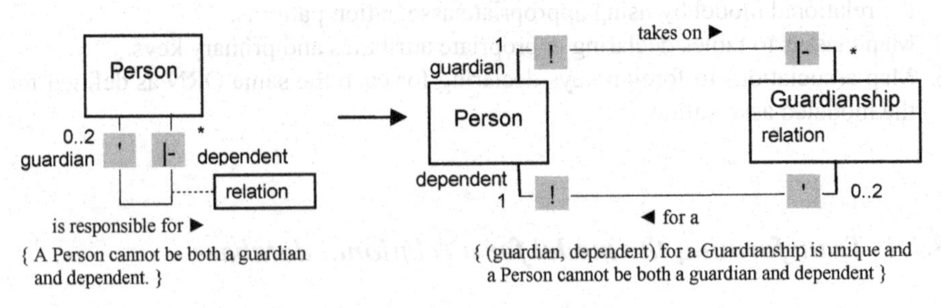

Fig. 8.2 Transformation of a many-to-many, intra-class association

(The "is responsible for" association in Fig. 8.2 can be seen as an intra-class association example of a Case 2 "is recorded for" pattern, given in Section 4.2.2.2, where the primary class and supporting class are the same. Here, deleting a guardian implicitly deletes all dependents who do not have a second guardian. A dependent Person object or a Guardianship object cannot be explicitly deleted. A Guardianship object is deleted by deleting its related guardian Person object or by explicitly destroying its link to this object.)

Fig. 8.3 shows the transformation of a ternary association, the classic association where vendors supply parts to projects. Many vendors can supply a part to a project, many parts can be supplied by a vendor to a project, and many projects can be supplied a part by a vendor—thus the many-to-many-to-many association.

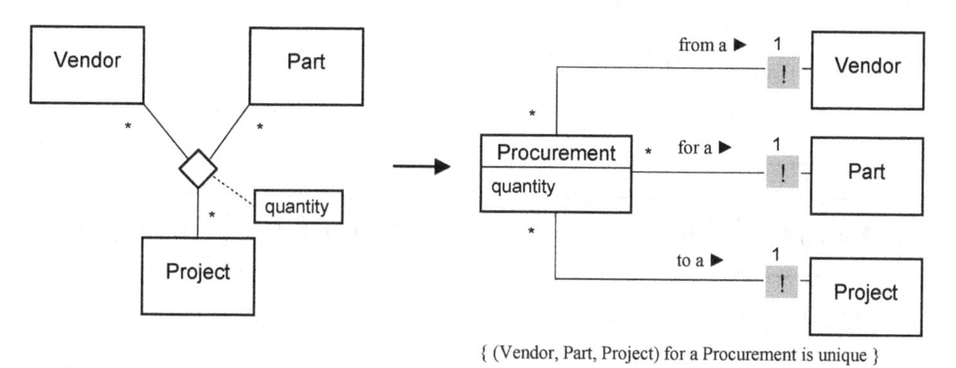

Fig. 8.3 Transformation of a ternary association

In an n-ary association where $n \geq 3$, it is possible, though rare, that one of the related class multiplicities is 1. For example, in Fig. 8.3 the multiplicity for Vendor could be 1 meaning that given a particular part and project, only one vendor is allowed to supply the part to the project. In this case, the mapping would not result in a uniqueness constraint that included the related class with multiplicity 1. In Fig 8.3, if the multiplicity for Vendor was 1 in the original model, the uniqueness constraint in the resultant model would be { (Part, Project) for Procurement is unique }.

We note that an associating class represents a weak entity type. Each associating object is dependent on each of the related objects for its existence and often its identity. This will be important later when we map an associating class to a table.

8.1.1.2 Association attributes for one-to-many and one-to-one associations

Remove association classes and association attributes for one-to-many associations by placing association attributes into the class on the many side of the association. For example, if startingDate is shown as an association attribute for a many-to-one, "works for" association between classes Employee and Unit, place this attribute into the Employee class.

For one-to-one associations, place association attributes into one of the classes, usually the class that is likely to have the fewest instantiated objects.

8.1.1.3 Generalization/specialization relationships

Using an appropriate "is a" association pattern, given in Section 4.2.7, transform each generalization/specialization relationship into an appropriate one-to-one association between the subclass and superclass. (This assumes we are implementing the model as a relational database, not an object-relational database.) For example, transform the two generalization/specialization relationships shown in Fig. 8.4 into the two one-to-one, "is a" associations shown in Fig. 4.15 (b). Likewise, transform the "is a" relationships shown in Fig. 8.5 into the associations shown in Fig. 4.16 (b).

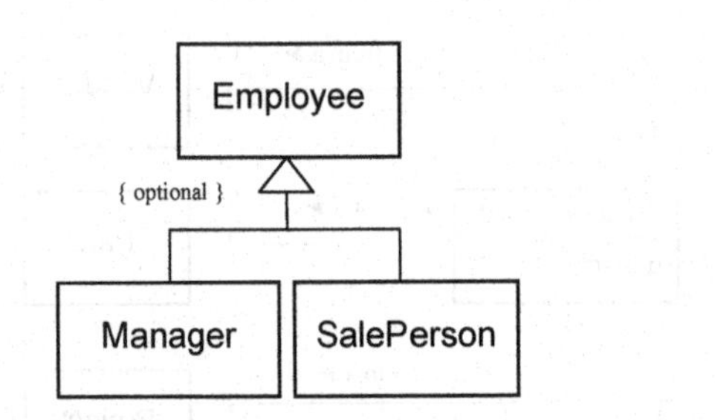

Fig. 8.4 Optional "is a" relationships

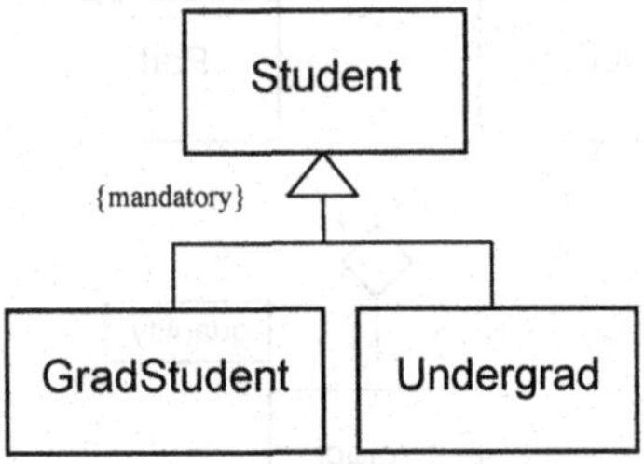

Fig. 8.5 Mandatory "is a" relationships

A subclass in the "is a" association represents a weak entity type. The subclass object is dependent on its related superclass object for its existence and identity.

8.1.1.4 Multi-valued attributes

Using the Case 1 "is recorded for" pattern, given in Section 4.2.2.1, transform each multi-valued attribute in a class into an appropriate supporting class and a one-to-many association. For example, as shown in Fig. 8.6, transform the attribute dependent[12] in a class Employee into the supporting class Dependent and the one-to-many "supports" association. This association is similar to the "claims" (or inverse "is recorded for") association given in Fig. 4.2 (b). Each element of dependent is a string representing the name of an employee's dependent.

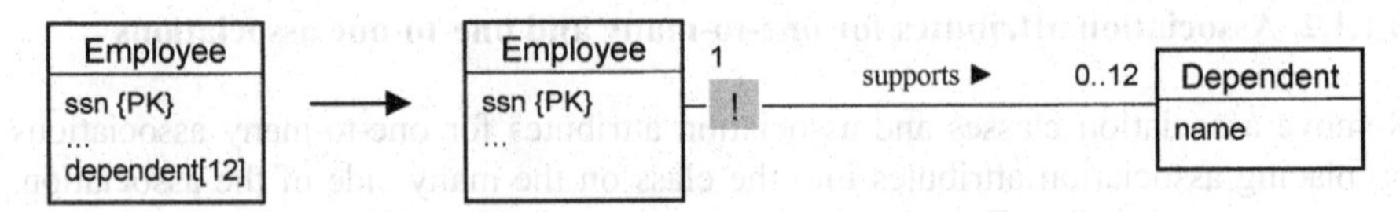

Fig. 8.6 Transforming a multi-valued attribute into a one-to-many association

As shown in Fig. 8.7, transform the attribute prevAddr[3] in the class Applicant into the supporting class prevAddr and one-to-many association "has lived at." Each prevAddr element is a string representing an applicant's previous address.

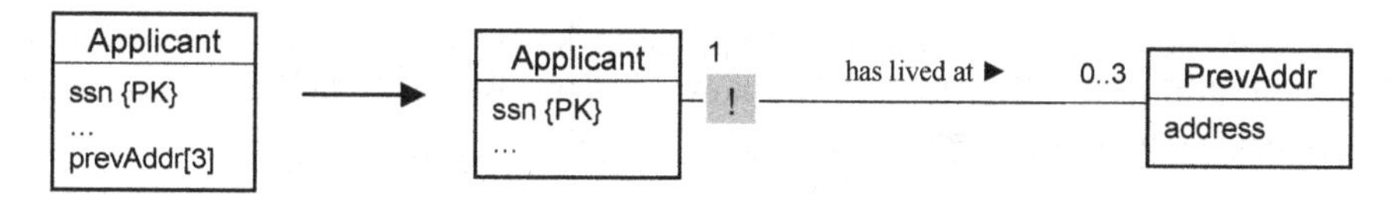

Fig. 8.7 Transforming a multi-valued attribute to a one-to-many association

The supporting classes created by the types of transformation shown above are always weak entity types.

8.1.2 Mapping classes to tables

After transforming the database model, we map each class to a table definition. More specifically, for each class create a table that includes all attributes of the class. For any composite attributes, include only the constituent simple attributes. Define for each attribute an appropriate type. If the class represents a weak entity type, include one or more primary key attributes from each *owner* class—e.g., an associated class, superclass, or primary class—to uniquely identify the rows of the created table. Declare a primary key for the relation.

Fig. 8.8 shows how classes Employee, a *strong entity type*, and Manager, a weak entity type, both modeled in Fig. 4.15 (b), are mapped to Employee and Manager tables, respectively. It also shows how class Dependent, modeled in Fig. 8.6 as a weak entity type, is mapped to table Dependent. Here, the ssn from the Employee table is included in Manager and Dependent tables to form primary keys.

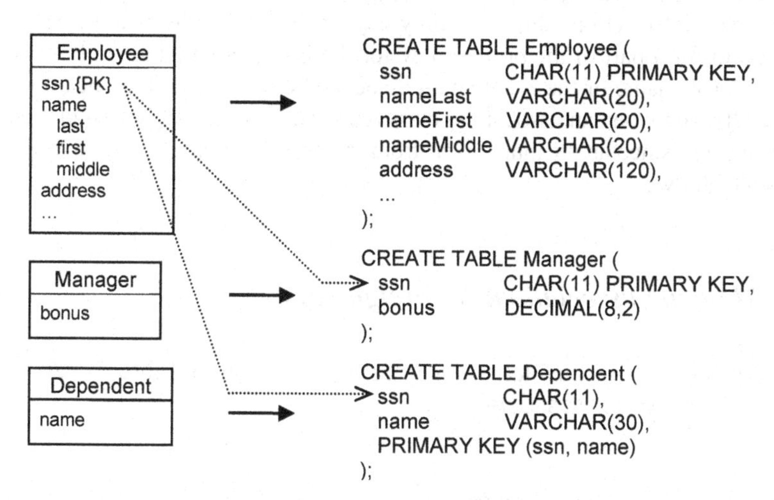

Fig. 8.8 Mapping classes representing strong and weak entity types to tables

Fig. 8.9 shows how the Guardianship class, shown in Fig. 8.2, is mapped to a Guardianship table. Fig. 8.10 shows how the Procurement class, shown in Fig. 8.3, is mapped to a Procurement table.

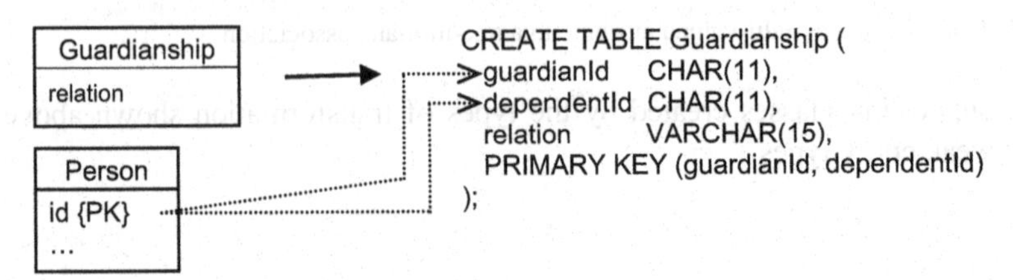

Fig. 8.9 Mapping the Guardianship class to a Guardianship table

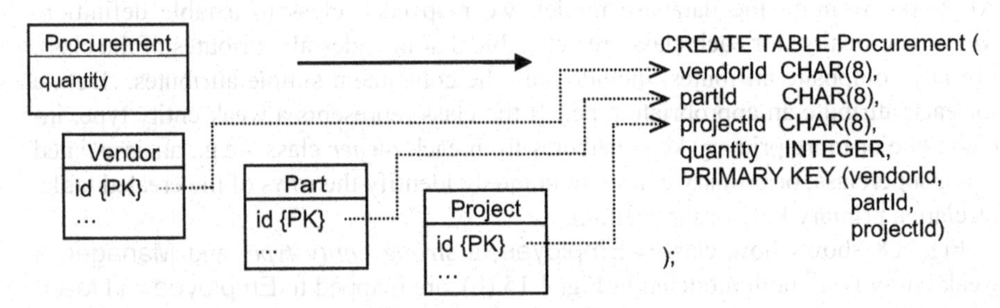

Fig. 8.10 Mapping the Procurement class to a Procurement table

A table that implements an associating class, like Guardianship or Procurement, is sometimes called an *intersection table*. When a weak, associating class is mapped to an intersection table, a primary key attribute taken from an owner, associated class to form part of the primary key for the intersecting table is often renamed to reflect a role name or the role that an associated object plays in association. This is especially true when an associating class represents an intra-class association, e.g., Guardianship, since an attribute taken from the associated class, e.g., id in Person, must be taken twice.

8.1.3 Mapping associations to foreign keys

After mapping classes to tables, we map each association to a foreign key declaration, which includes an <association>.

8.1.3.1 One-to-many associations

For each one-to-many association, place a copy of the primary key attribute(s) of the table representing the class (or role) at the *one-end* of the association into the table representing the class (or role) at its *many-end* to act as a foreign key. Do not include an attribute if it is already present in the table (for instance, if already included by step 2). If appropriate and not yet done, rename the foreign key attribute(s) to reflect the role that the referenced object plays in the association.

Declare the foreign key contraint, and define for it the many-to-one, syntactic *<association>* corresponding to its graphical representation in the database model.

Fig. 8.11 shows how the one-to-many association between units and employees, given in Fig. 6.4, is mapped to a foreign key declaration. The declaration is given using T-SQL/ORN Additive and must be a named foreign key constraint. Integration of ORN into the SQL standard, as discussed in Chapter 11, would require only an "unnamed" REFERENCES clause that includes the *<association>*.

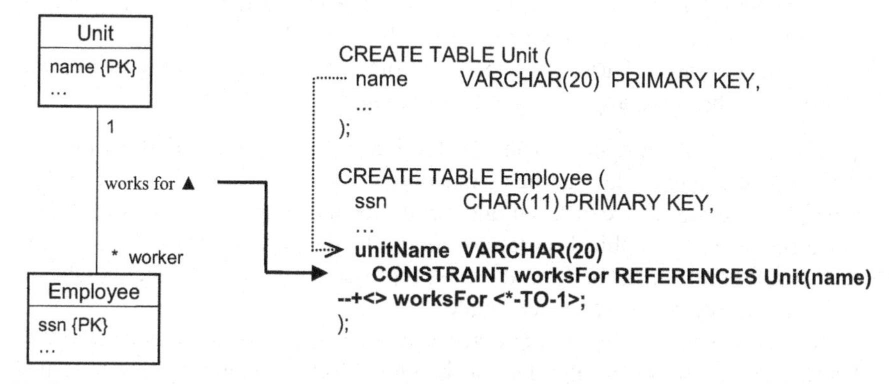

Fig. 8.11 Mapping a one-to-many association to a foreign key

Fig. 8.12 shows how the two one-to-many associations between classes Person and Guardianship, given in Figs. 8.2 and 8.9, are mapped to foreign key declarations. Here, since Guardianship is a weak entity type, the needed ids from the table at the one-end, i.e., Person, have already been included into the Guardianship table. This was done in Step 2 of the mapping process (see Fig. 8.9).

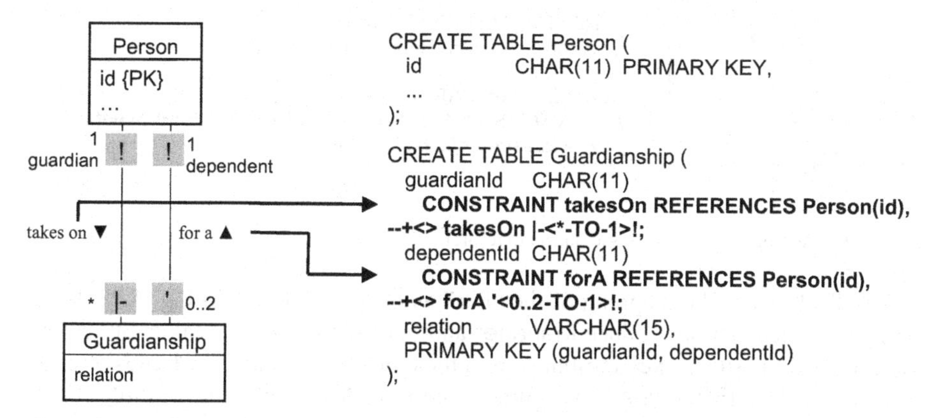

Fig. 8.12 Mapping two one-to-many associations to foreign keys

8.1.3.2 One-to-one associations

For each one-to-one association, place a copy of the primary key attribute(s) of the table representing the class (or role) at one end of the association into the table representing the class (or role) at the other end to act as a foreign key. (We assume here that there is no reason to combine these tables, an option when the association is not intra-class.) Do not include an attribute if it is already present in the table (for instance, if already included by step 2). Use the following guidelines to select the table that will contain the foreign key.

- If the association is 1-to-1, select the table having the largest primary key since this results in the smallest foreign key and thus less storage.
- If the association is 1-to-0..1, select the table representing the 0..1 end of the association since this results in no foreign keys that are null.
- If the association is 0..1-to-0..1, select the table likely to have the fewest number of rows since this generally results in less storage.

If appropriate and not yet done, rename the foreign key attribute(s) to reflect the role that the referenced object plays in the association.

Declare the foreign key constraint and define for it the syntactic <*association*> that corresponds to its graphical representation in the database model. The direction of this <*association*> must be from the class represented by the referencing table to the class represented by the referenced table.

Fig. 8.13 shows how a one-to-one association between company credit cards and employees is mapped to a foreign key declaration. Very few employees are issued company credit cards. An employee cannot be deleted without first "retrieving the card" and explicitly destroying the "is issued" link, i.e. setting cardHolder to NULL.

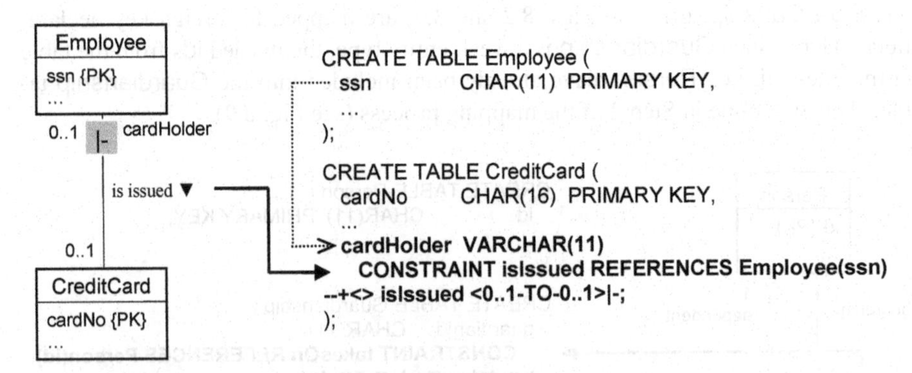

Fig. 8.13 Mapping a one-to-one association to a foreign key

Fig. 8.14 shows the mapping of the "is a" associations shown in Fig. 4.15 (b) to foreign key declarations. Since Manager and SalesPerson are weak entity types, the needed ssn primary key attributes from the table representing the 1 ends of each association, i.e., Employee, have already been included into the Manager and SalesPerson tables. This was done in Step 2 of the mapping process (see Fig. 8.8).

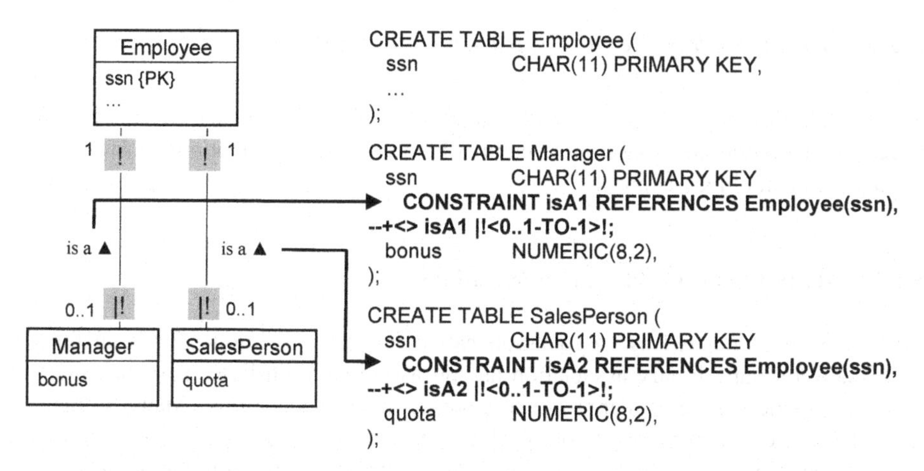

Fig. 8.14 Mapping two one-to-one, "is a" associations to foreign keys

8.1.3.3 Ensuring that links can be explicitly destroyed

When a foreign key is part of a primary key, it cannot be set to null. This means that an association link, as represented by a non-null foreign key value, cannot be explicitly destroyed. If such link destruction is required, then another primary key must be declared for the referencing table. This could be a different candidate key or an IDENTITY column.

8.2 Mapping an ORN-Extended Model to an Object DDL

This section provides directions for mapping a database model given by an ORN-extended class diagram to an ORN-extended object DDL. Here, the examples given use ODDL, which was discussed in the previous chapter. Chapter 12 describes another ORN-extended object DDL, namely the Object Definition Language (ODL), which is defined by the ODMG 3.0 standard (Cattel et al. 2000). The directions below also apply to ODL; the ODDL given in examples can be easily translated.

The basic steps in the mapping process, which are given in greater detail in the subsections that follow, are:

1. Transform the database model, if necessary, into a model more compatible with an object database model by using appropriate association patterns.
2. Map classes in the model to class definitions and extents in the object DDL declaring appropriate value-based attributes for each class.
3. Map associations to object-based attributes declaring for each the same ORN as defined for the modeled association.

8.2.1 Transforming the model for an object database

To simplify the mapping of the class diagram model to the object DDL, we first transform the model to make it more compatible to the way associations must be implemented in an object database.

8.2.1.1 Many-to-many and *n*-ary associations

Using the "is associated by" association pattern, given in Section 4.2.4, transform each many-to-many binary association having association attributes and each *n*-ary association where $n \geq 3$ into appropriate *n* one-to-many associations. Many-to-many associations not having association attributes may also be transformed into one-to-many associations to facilitate future association attributes. The directions for doing these transformations are given in Section 8.1.1.1 and are not repeated here except for providing the following example, which slightly revises the Employee-Project example given in Section 8.1.1.1, Fig. 8.1.

Fig. 8.15 shows the transformation of the many-to-many association between classes Employee and Project into an associating class Assignment; two one-to-many, "is associated by" association types; and an appropriate constraint.

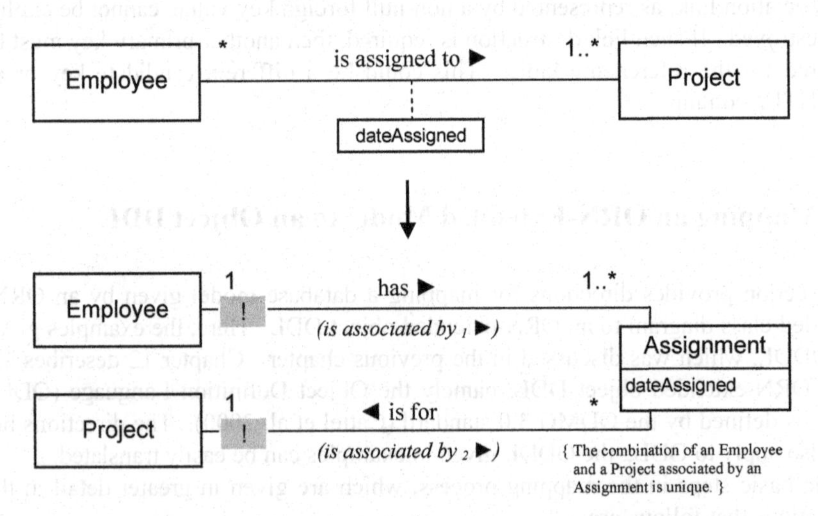

Fig. 8.15 Many-to-many association transformed into two one-to-many associations

8.2.1.2 Association attributes for one-to-many and one-to-one associations

Remove association classes and association attributes for one-to-many associations by placing association attributes into the class on the many side of the association.

For example, if startingDate is shown as an association attribute for a many-to-one, "works for" association between classes Employee and Unit, place this attribute into the Employee class.

For one-to-one associations, place association attributes into one of the classes, usually the class that is likely to have the fewest instantiated objects.

These transformations are the same as those described in Section 8.1.1.2.

8.2.2 Mapping classes to class definitions and extents

After transforming the database model, we map each class to a class definition. More specifically, for each class create a class definition that includes all attributes of the class. If the class is a subclass, include the appropriate isa specification to identify its superclass.

If the database objects (or *persistent objects*) of a class must be accessed directly, create an extent for the class. A class that represents a strong entity type usually requires an extent. An appropriate collection type may be specified for the extent. Although not currently implemented in ODDL, some collection types allow keys to be specified, e.g., Dictionary in the ODL of ODMG 3.0. Primary keys may be implemented in this manner.

Fig. 8.16 shows how classes Employee, a strong entity type; Manager, a subclass; and Dependent, a weak entity type, are mapped to ODDL. Composite attributes, like name, and multi-valued attributes, like dependent in Fig. 8.6, can be defined as C++ user-defined types. Here, however, the attribute dependent has been transformed into the class Dependent and a one-to-many association.

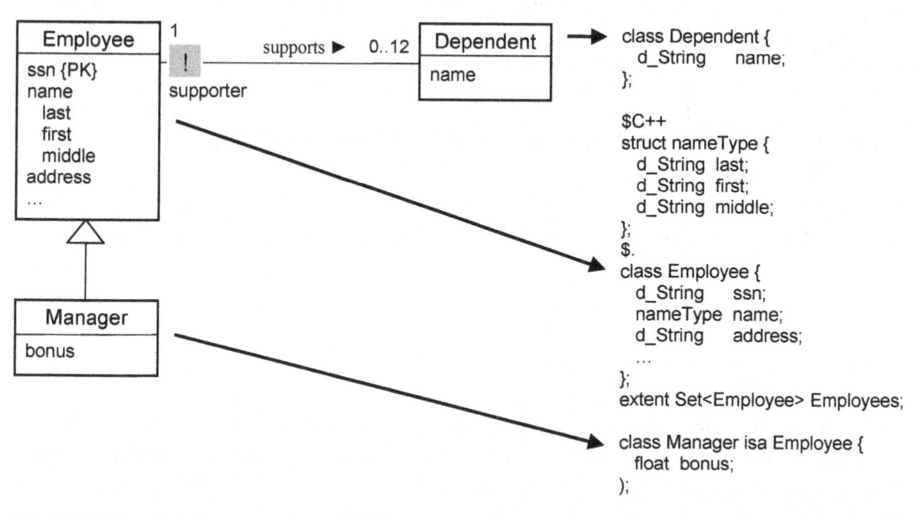

Fig. 8.16 Mapping classes to class definitions and extents

8.2.3 Mapping associations to object-based attributes

After mapping classes to class definitions and extents, we map each association to a pair of object-based attribute declarations, one of which includes an <association>. We first focus on inter-class associations, then address intra-class associations.

8.2.3.1 One-to-one associations

For each one-to-one association between classes *A* and *B*, add a single-valued, object-based attribute (a reference) to class *A* to reference an object of type *B* and a single-valued, object-based attribute to class *B* to reference an object of type *A*. If appropriate, name the attributes to reflect the role that the referenced (related) object plays in the association. Define the attributes as inverses of each other.

Define for one of the attributes the one-to-one, syntactic <association> that corresponds to its graphical representation in the database model. The direction of this <association> is from the referencing class to the referenced class.

Fig. 8.17 shows how a one-to-one association between employees and company credit cards is mapped to a pair of references. (An employee [Employee object] cannot be deleted without first "retrieving the card" from the employee and explicitly destroying the "is issued" link, i.e., setting creditCard [or cardHolder] to 0.)

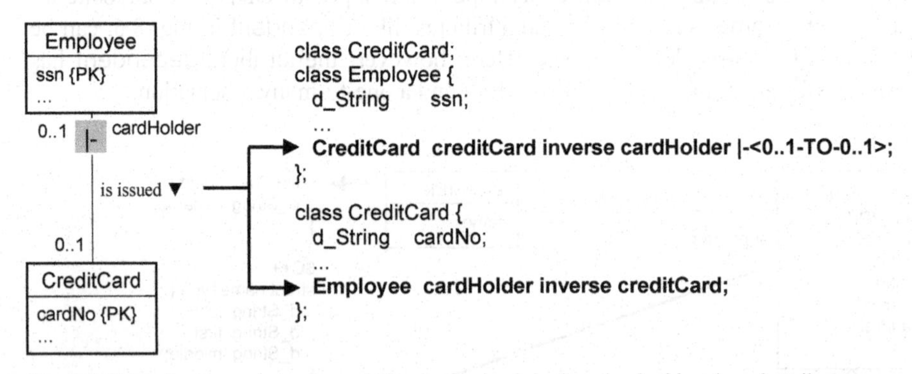

Fig. 8.17 Mapping a one-to-one association to a pair of single-valued, object-based attributes

8.2.3.2 One-to-many associations

For each one-to-many association between classes *A* and *B*, add a multi-valued, object-based attribute (a collection of references) to class *A* to reference objects of type *B* and a single-valued, object-based attribute (a reference) to class *B* to reference an object of type *A*. Define the desired type, such as a Set<*B*>, for the collection. If appropriate, name the attributes to reflect the role that the referenced (related) object(s) play(s) in the association. Declare the attributes as inverses of each other.

Define for the collection attribute the one-to-many (or for the reference attribute the many-to-one), syntactic *<association>* that corresponds to its graphical representation in the database model.

Fig. 8.18 shows how a one-to-many association between employees and dependents (shown previously in Fig. 8.16) is mapped to a pair of object-based attributes. (If an Employee object is deleted, all Dependent objects referenced by the dependents are implicitly deleted. Also, if a Dependent object reference is removed from dependents [or if supporter in a Dependent object is set to 0], the Dependent object is implicitly deleted.)

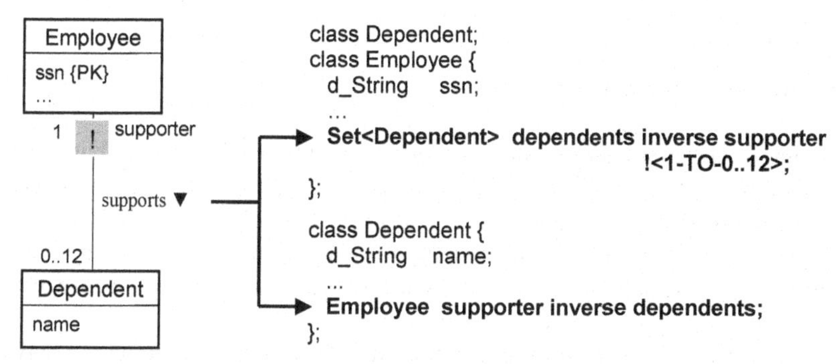

Fig. 8.18 Mapping a one-to-many association to a pair of object-based attributes

8.2.3.3 Many-to-many associations

For each many-to-many association between classes *A* and *B*, add a multi-valued, object-based attribute (a collection of references) to class *A* to reference objects of type *B* and a multi-valued, object-based attribute (a collection of references) to class *B* to reference objects of type *A*. Define the desired types, such as a Set<*B*> and a List<*A*>, for the collections. If appropriate, name the attributes to reflect the role that the referenced (related) objects play in the association. Declare the attributes as inverses of each other.

Define for one of the attributes the many-to-many, syntactic *<association>* that corresponds to its graphical representation in the database model. The direction of this *<association>* is from the referencing class to the referenced class.

Fig. 8.19 shows how a many-to-many association between employees and projects is mapped to a pair of object-based attributes. Unlike the employees and projects association shown in Fig. 8.15, this association has no association attribute, and it can never have association attributes if modeled and defined as shown in Fig. 8.19.

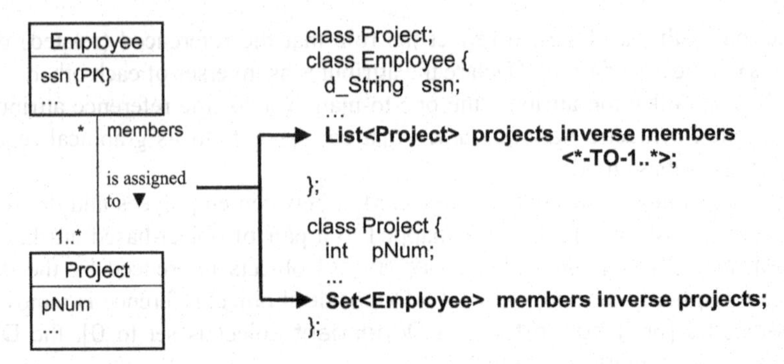

Fig. 8.19 Mapping a many-to-many association to a pair of multi-valued, object-based attributes

8.2.3.4 Intra-class associations

For each intra-class association within a class *A*, add appropriate object-based attributes to reference related objects of type *A* as prescribed for inter-class associations treating each end of the association as if *A* was a distinct class. Two caveats to this directive are 1) for each one-to-one or many-to-many association that is symmetric, add only one object-based attribute making it an inverse of itself and 2) **always** use appropriate role names to name the attributes.

Fig. 8.20 shows how a one-to-many, intra-class, "is a part of" association between parts is mapped to a pair of object-based attributes. (If a Part object is deleted, all Part objects referenced by the attribute components are implicitly deleted.)

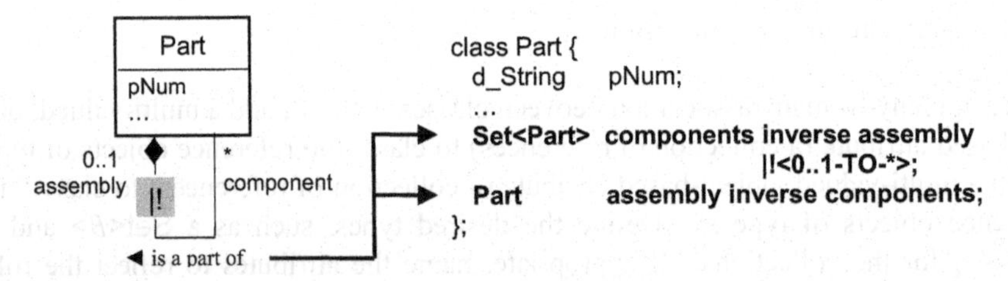

Fig. 8.20 Mapping a one-to-many, intra-class association to a pair of object-based attributes

Fig. 8.21 shows how a many-to-many, intra-class association between people is mapped to a pair of object-based attributes. Unlike the "is responsible for" association shown in Fig. 8.2, this association has no association attribute, and it can never have association attributes if modeled and defined as shown in Fig. 8.21. (If a Person object is deleted, any Person object referenced by the attribute dependents is implicitly deleted unless the guardians attribute in this object references another Person object. When a reference to a Person object is removed from the attribute dependents, the referenced Person object is implicitly deleted unless the guardians attribute in this object references another Person object. A Person object cannot be deleted if its guardian attribute contains an **existing** reference.)

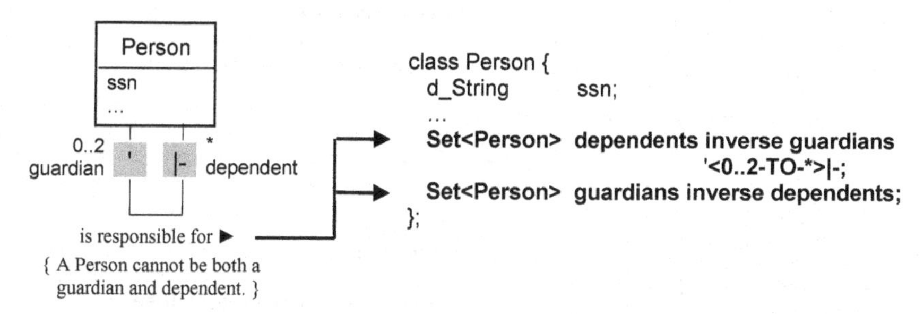

Fig. 8.21 Mapping a many-to-many, intra-class association to a pair of object-based attributes

Fig. 8.22 shows how another many-to-many, intra-class association between people is mapped to an object-based attribute. Only one attribute is required since the association is symmetric.

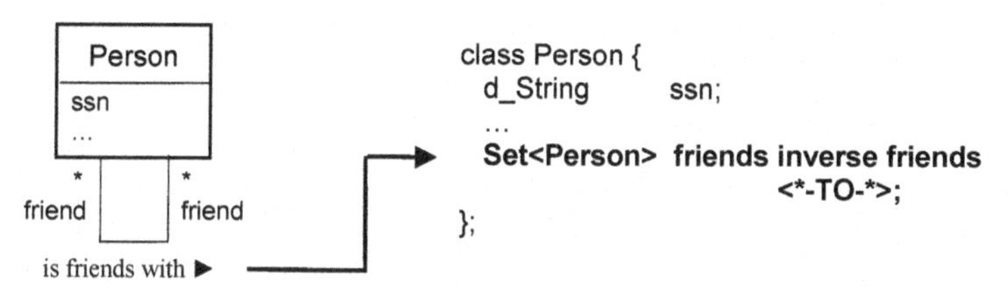

Fig. 8.22 Mapping a many-to-many, intra-class, symmetric association to a single object-based attribute

Fig. 8.23, on the following page, shows how two different one-to-one, intra-class associations between people, which both record marriages, are mapped to object-based attributes. It provides another example showing that the symmetry of an association affects the number of object-based attributes required. Fig. 8.23 (a) models the association as non-symmetric and shows its mapping. Fig. 8.23 (b) models the association as symmetric and shows its mapping.

8.3 Conclusion

As shown in this chapter, the most difficult tasks in mapping associations from a database model to a DDL are transforming the model and mapping the associations into their logical structural representations—foreign keys or object-based attributes. The easy task, when both the model and DDL include ORN, is mapping the association semantics.

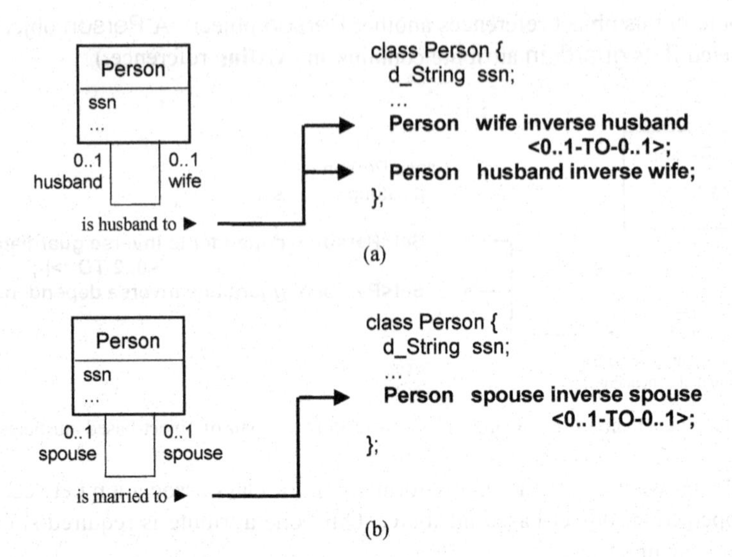

Fig. 8.23 Mapping a one-to-one, intra-class association that is modeled (a) as non-symmetric and (b) as symmetric

Chapter 9

Association Semantics

Dealing with the Subtleties, Inconsistencies, and Ambiguities

ORN allows database developers to better document association semantics during system analysis and design and thus recognize and address semantic subtleties. These subtleties can lead to associations and combinations of associations that may be mathematically inconsistent or ambiguous. With a modeling tool, like the ORN Simulator, association semantics can be fine-tuned and any inconsistencies or potential ambiguities can be detected by database developers with experimentation and by the tool itself. Thus developers need not design, code, and test complex code and/or triggers before flaws in associations are detected. And of course with ORN, once association semantics have been confirmed with modeling, they can be easily mapped to an ORN-extended DDL and automatically maintained by the DBMS.

This chapter discusses some of the inconsistencies and ambiguities in association semantics of which a database developer should be aware. Such inconsistencies and ambiguities can be easily identified when associations are specified using ORN. This is not the case when they are implemented in application programs, class methods, or SQL constraints and triggers. The chapter revises and extends a previous paper on this subject, Hardeman and Ehlmann (1996).[1] Section 9.1 of the chapter describes a number of association semantic inconsistencies. Section 9.2 discusses the possible detection of these inconsistencies in modeling tools and DDLs. Section 9.3 presents associations as functions to mathematically prove the intra-class inconsistencies discussed in Section 9.1. This section can be skipped by the practical database developer. Section 9.4 discusses association semantic ambiguities, which are only identified and briefly described here but are more fully examined in the next chapter. Section 9.5 provides concluding remarks.

9.1 Inconsistencies

Inconsistencies in association semantics can occur in both *intra-class* and *inter-class associations*. In the latter associations the subject class and relative class are different, while in the former associations they are the same. Association inconsistencies are detected by both ODDL and the ORN Simulator, and appropriate messages are displayed. Some inconsistencies occur within the specification of a single *<association>* while others involve *<association>* combinations sharing a common class.

---

[1] This work is based on an earlier work: "Relationship behavior in object databases: subtleties and inconsistencies," in *Proc. ACM Southeast Regional Conf.*, 0-89791-826-6 © ACM, 1996.

B.K. Ehlmann, *Object Relationship Notation (ORN) for Database Applications*,
Advances in Database Systems 39, DOI 10.1007/978-0-387-09554-7_9,

To describe both types of inconsistencies, some variables are needed. Let us assume an association A between a subject class S and related class R. An <association> specification for this association can be described as

$$b_{SI}b_{SX}<m_{SL}..m_{SU}\text{-to-}m_{RL}..m_{RU}>b_{RX}b_{RI}$$

where b_{SI} is the subject class implicit binding, b_{SX} is the subject class explicit binding, m_{SL} is the subject class lower bound, m_{SU} is the subject class upper bound, m_{RL} is the related class lower bound, m_{RU} is the related class upper bound, b_{RX} is the related class explicit binding, and b_{RI} is the related class implicit binding. In using these variables to describe associations we must be mindful of the following:

- The class R may be class S with related objects playing a different role in an intra-class association.
- The values of the binding variables may be explicitly or implicitly specified in an <association>. For example, !<1-to-*>|- explicitly specifies that b_{RI} = |- and implicitly specifies that b_{SI} = |!, b_{SX} = |!, and b_{RX} = nil.
- $m_{SU} > m_{SL}$ and $m_{RU} > m_{RL}$ when m_{SU} = * and m_{RU} = *, respectively.
- Where two associations are being considered, the variables must be qualified, e.g., $A.m_{SL}$ and $A'.m_{SL}$.
- Inconsistent associations are described from the subject class perspective. Thus, even though an association A may appear consistent from this perspective, its inverse A^{-1}, i.e., the association from the perspective of the related class being the subject class, may be inconsistent.

9.1.1 Within an <association>

Here, inconsistencies are described that can be identified by simply examining a single <association>.

9.1.1.1 Involving intra-class associations

The condition IA_1 (*I*nconsistent *A*ssociation $_1$) given below describes an intra-class association that is inconsistent—actually, mathematically invalid and thus impossible to realize.

IA_1: ((m_{SL} = 0 and m_{SU} = 1) and $m_{RL} \geq 1$) or
((m_{SL} = 1 and m_{SU} = 1) and $m_{RL} \geq 1$ and $m_{RU} > 1$)

The inconsistencies of the intra-class associations 0..1-to-1, 0..1-to-1..*, and 1-to-1..* were formally proven in Ehlmann (1992). These proofs are given in Section 9.5, where associations are interpreted as functions; however, the reader may wish to skip the formality of this section and settle for an informal proof on the inconsistency of just one of these associations.

We can easily show that a 0..1-to-1 intra-class association is impossible. Assume such a one-to-one "is a mentee of" association between *n* employees where each employee must have a mentor, i.e., from the perspective of the "mentee" the association is -to-1. There must be *n* mentors since no two employees can have the same mentor (otherwise, a mentor would have two mentees, and the relationship would be many-to-one). Since every employee is a mentor, then every employee must have a mentee. Therefore, the association must be 1-to-1 and cannot be 0..1-to-1.

It is somewhat surprising that some textbooks on database management actually discuss how such associations should be implemented, even though they are impossible! (See, for example, p. 467 of Connolly and Begg (2005).) Perhaps this is indicative of the overall lack of concern with and study of association semantics as was discussed in the Preface and Chapter 1 of this book.

9.1.1.2 Involving the ? binding

Another inconsistent association, whether an intra-class or inter-class association, is described by the condition IA_2, given below.

IA_2: ($b_{S|}$ = |? or b_{SX} = X?) and m_{SL}= 0

Such an *<association>*, for example ?<0..1-to-*>, is inconsistent because a ? binding specifies that on implicit (|?) or explicit (X?) link destruction the related object is implicitly deleted when m_{SL} is violated, but a multiplicity with a lower bound of 0 can never be violated on link destruction.

A related association, while not inconsistent, is nevertheless needlessly complex and a bit inefficient. It is described by the condition ($b_{S|}$ = |? or b_{SX} = X?) and m_{SL} = m_{SU}. Most often, the association is a ?<1-to-..., |?<1-to-..., or X?<1-to-...association. Here, the m_{SL} of 1 is always violated on implicit (|) or explicit (X) link destruction, so why make the system check for a violation? When m_{SL} = m_{SU}, a ! binding should be used instead of a ? binding.

9.1.1.3 Involving required deletion of a related object and its implicit bindings

The remaining association inconsistencies discussed in this subsection and the next are more esoteric and more difficult to describe. Two additional variables will be useful. The variable *SODrROD*, meaning Subject Object Deletion requires Related Object Deletion, is defined as:

SODrROD: $b_{S|}$ = |! or (m_{SL} = m_{SU} and ($b_{S|}$ = |? or $b_{S|}$ = |'))

This condition guarantees for an association that an implicit deletion of any related object is required when a subject object is deleted.

A similar variable, *ELDrROD*, meaning Explicit Link Destruction requires Related Object Deletion, is defined as:

ELDrROD: b_{SX} = X! or (m_{SL} = m_{SU} and (b_{SX} = X? or b_{SX} = X'))

This condition guarantees for an association that an implicit deletion of the related object is required when a link is explicitly destroyed.

Using the above variables, the remainder of this subsection and the next subsection give conditions that describe <*association*>s that are **likely** inconsistent. Since inconsistency is not a certainty, these associations should not be unequivocally disallowed. In certain, unusual circumstances, link cycles together with other associations can render normally inconsistent associations consistent. More specifically, deletion of a subject class object or the implicit deletions of related objects may cause the implicit deletion of other subject class objects, possibly negating a lower bound multiplicity violation or the processing of a |- binding that would have otherwise occurred. A simple example to illustrate such a situation is given later in this subsection. *Link cycles* occur when an object is indirectly related to itself and are discussed further in Section 9.3 and more fully in Chapter 10.

The condition IA_3 given below describes associations that are likely inconsistent because on deletion of a subject object or explicit destruction of an A link, a related R object must be implicitly deleted but cannot be deleted because of the related class multiplicity and/or implicit binding and the related object's A links with other subject class objects.

$$IA_3\text{: } (SODrROD \text{ or } ELDrROD) \text{ and } m_{SL} > 1 \text{ and } ((m_{RL} = m_{RU} \text{ and } b_{R|} = \text{nil}) \text{ or } b_{R|} = \text{|-})$$

The condition $m_{SL} > 1$ ensures that another A link involving the related R object exists besides the one being implicitly destroyed.

Some examples of such associations are given below.

|!<2..*-to-1> !<2..*-to-1>|- |?<3..3-to-0..1>|- '<2..2-to-3..3>

X!<2..*-to-1> !<2..*-to-0..1>|- X?<3-to-1> X'<2-to-*>|-

Fig. 9.1 shows a model and database that can be used to easily illustrate the inconsistency of the |!<2..*-to-1> association. The deletion of `x1` results in the destruction of the `x1↔y1` link and requires the deletion of `y1` (see Table 2.1). `y1` is deleted and the `x2↔y1` link is implicitly destroyed, but at the end of the complex object operation, a lower bound violation results since `x2` is not related to a `Y` object. So, the deletion of `x1` fails.

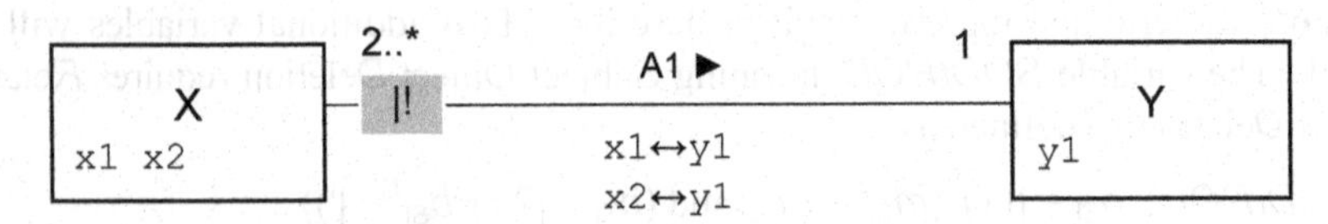

Fig. 9.1 Initial database for a !<2..*-to-1> association

Fig. 9.2 shows the same model and database as that shown in Fig. 9.1 except that another association A2 has been added along with an A2 link between objects `x1` and `x2`. The added association and link creates a link cycle. Now, the deletion of x1 results in the same scenario as that discussed for Fig. 9.1 except that the deletion of x1 also results in the implicit deletion of x2. So, at the end of the complex object operation, no lower bound violation results, and the deletion of x1 succeeds. Such cases are why a likely inconsistent association should not be disallowed.

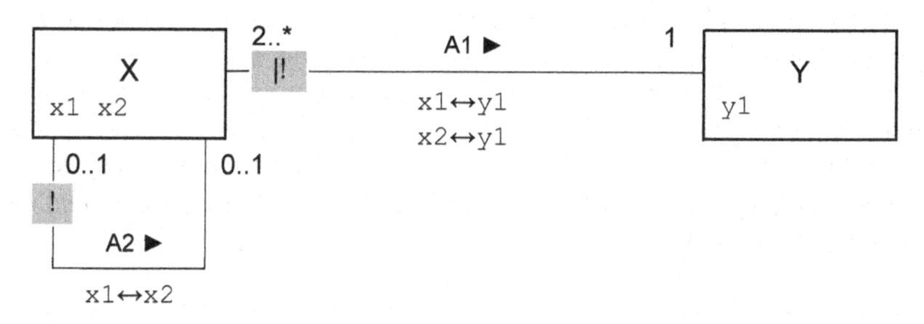

Fig. 9.2 Initial database for a !<2..*-to-1> association with link cycle y1↔x1, x1↔x2, x2↔y1

9.1.1.4 Involving required deletion of a related object and its explicit bindings

The condition IA_4 given below describes associations that are likely inconsistent because on explicit destruction of an *A* link, a related *R* object must be implicitly deleted but cannot be deleted because of the related class multiplicity and/or explicit binding.

IA_4: *ELDrROD* and (($m_{RL} = m_{RU}$ and b_{RX} = nil) or b_{RX} = X-)

Some examples of such associations are given below.

X!<1-to-1> !<1-to-0..1>X- X?<3..3-to-1> '<1-to-3..3>

Although explicit link destruction may be impossible with these associations, link changes involving replacement of the related object may still be valid, which is another reason for not disallowing them. For example, assume the model and database shown in Fig. 9.3. Attempting to destroy the link x1↔y1 results in a lower bound multiplicity exception since y1 would be implicitly deleted (because of X!) and thus x1 would not be linked to a Y object. (The link x1↔y1, however, can be changed in a single operation that replaces y1 with another Y object, such as y2, an object created within an as yet uncommitted transaction. The y1 object is implicitly destroyed and no exception results since x1 has its required Y object, y2).

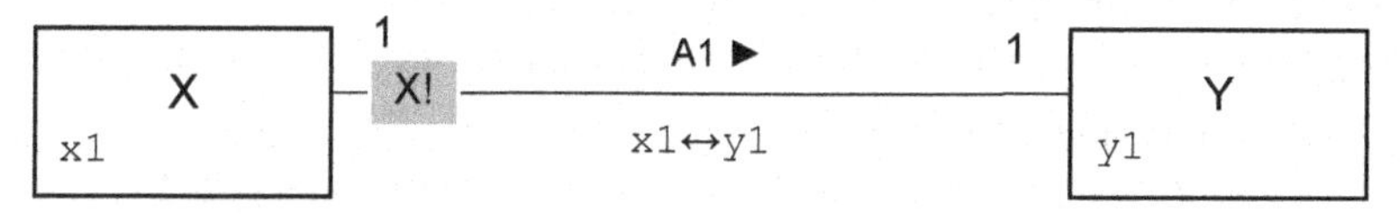

Fig. 9.3 Initial database for a X!<1-to-1> association

9.1.2 Involving <association> combinations

Inter-class association inconsistencies related to bindings occur when the deletion of an object is made impossible due to its association with a related class object and the related object's association with another object. For example, an inconsistency occurs when an association between classes X and Y is defined as |!<1-to-1> and another between classes Y and Z is defined as <1-to-1>. This situation is shown in Fig. 9.4. Here, deletion of an X object, e.g., `x1`, is made impossible since it requires implicit deletion of the related Y object, e.g., `y1`, but deletion of a Y object requires destroying its link with a Z object, e.g., `z1`, but this violates the 1 multiplicity for Y in the association A2 and is disallowed by the default implicit destructibility binding for class Y in the A2 association.

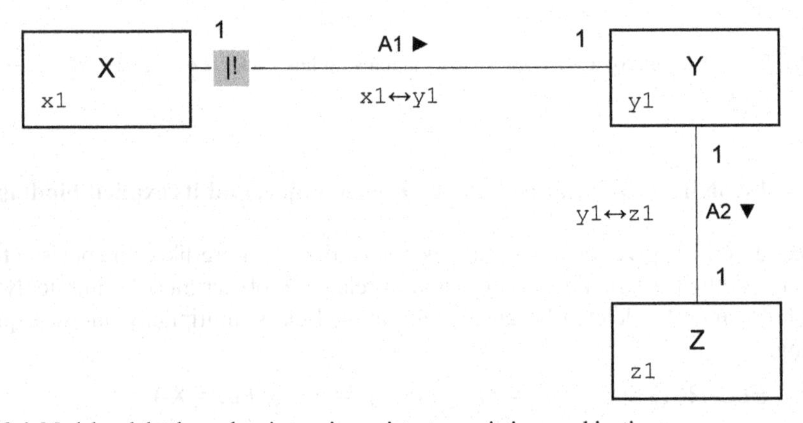

Fig. 9.4 Model and database showing an inconsistent association combination

To generically describe such inconsistent association combinations, we again assume the variables given in the previous subsection for describing an *<association>* and assume an association *A* between a subject class *S* and related class *R*. Furthermore, we assume that *A'* is an association between class *R* and another class (or role), which could be *S*, and that *A'* is not A^{-1}, the inverse of *A*.

The condition IA_5 given below describes combinations of *A* and *A'* that are likely inconsistent because on deletion of an *S* object or explicit destruction of an *A* link, a related *R* object must be implicitly deleted but cannot be deleted because of the bindings and multiplicities of *A'*.

$$IA_5\text{: } (A.SODrROD \text{ or } A.ELDrROD) \text{ and } A'.m_{RL} > 0 \text{ and } ((A'.m_{SL} = A'.m_{SU} \text{ and } A'.b_{S|} = \text{nil}) \text{ or } A'.b_{S|} = |\text{-})$$

$A'.m_{RL} > 0$ ensures that an *A'* link involving the related *R* object exists.

Some examples of such associations are given below, the first shown in Fig. 9.4.

|!<1-to-1> and <1-to-1>
'<1-to-2..*>|- and |-<0..1-to-1..*>
X!<1-to-*>|- and |-<0..1-to-2..*>
|!<1-to-*> and <1-to-1..*>
?<3..3-to-1> and <1-to-1>
?<3..3-to-*> and <1-to-1>

9.2 Inconsistency Detection

The algorithm given in Fig. 9.5 verifies the consistency of ORN specifications given for all associations in a database. It has been implemented in C++ and integrated into the analysis phase of the ODDL processor of OR+. It traverses through all association descriptions in the meta-database analyzing multiplicities and bindings. When an inconsistency is detected, an appropriate error or warning message is output.

```
for each ORN-related class C
  for each association A where C is the subject class
    Let S = C and R = related class of C in A

    // Check for inconsistencies within A
    if R = S then
      if A.IA1 then Display an invalid intra-class association error for A
    if A.IA2 then Display an invalid use of ? binding error for A
    if A.IA3 then Display an appropriate inconsistent association warning for A
    if A.IA4 then Display an appropriate inconsistent association warning for A

    // Check for inconsistencies involving A and other associations
    for each A' where R is the subject class and A' ≠ A^-1
      if IA5 then Display an appropriate inconsistent association warning for A and A'
```

Fig. 9.5 Algorithm for detecting association inconsistencies within a database model or definition

The algorithm results in all associations being checked for inconsistencies from the perspective of each class (or role) serving as the subject class (or role). All combinations of associations sharing a common class are also checked for inconsistencies. The checks for inconsistencies within a single association can be done immediately after an *<association>* is specified (or parsed), while the checks for inconsistencies involving association combinations can be deferred until all *<association>*s have been specified.

As shown in Chapter 3, the ORN Simulator allows users to define associations using ORN and observe association semantics as objects are created and deleted and association links are created, destroyed, and changed. Association inconsistencies are detected by the algorithm given in Fig. 9.5 and messages that flag these inconsistencies as errors or warnings appear on the ORN Simulator screen when the user enters the *<association>* for a relationship. For example, the screen shown in Fig. 9.6 shows the warning message that appears after the user defines the *<association>* for the R2 relationship as 1-to-1. Here, the ORN Simulator has detected the inconsistent combination of associations that was discussed in Section 9.1.2 and illustrated using Fig. 9.4.

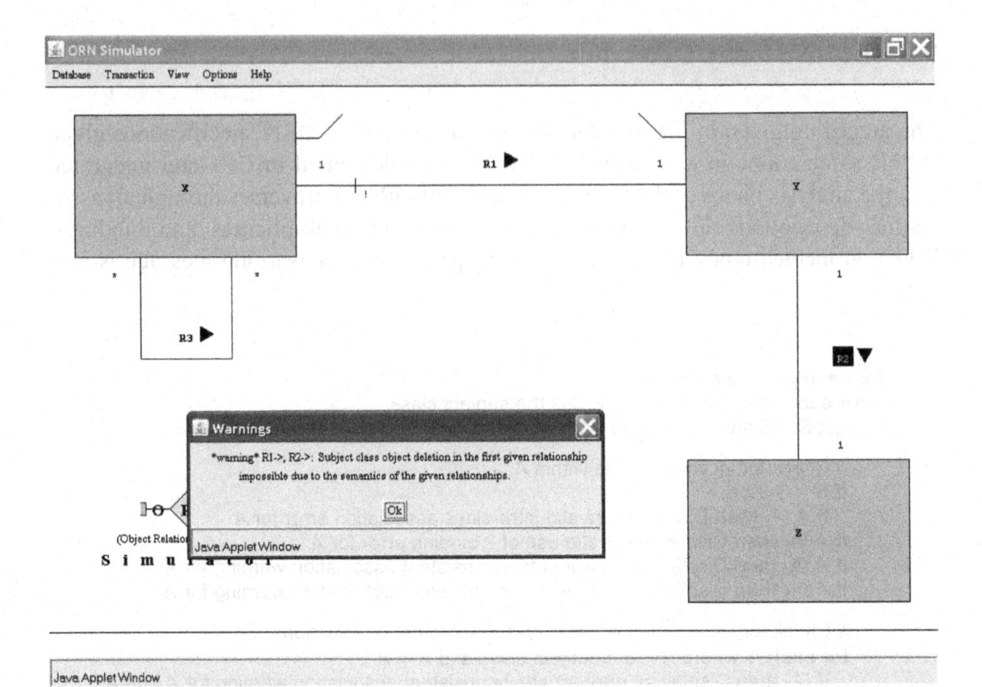

Fig. 9.6 Screen showing detection of association inconsistency in the ORN Simulator.

9.3 Ambiguities

Association semantics, though consistent, can sometimes result in ambiguities in terms of what the final outcome will be after object deletions, link destructions, or link changes. Such ambiguities occur because of the recursive nature of most association semantic specifications, because of link cycles within a database, and because the ordering in which links are processed by a database system is often indeterminate. In SQL, association ambiguities may occur when the ON DELETE RESTRICT referential action is specified (Date 1990). In ORN, association ambiguities may occur with two types of specifications. These are covered extensively in Chapter 9, here they are just identified and briefly described.

The first specification is a *cyclic one-ended* |-. Using the variables defined in the previous sections for describing an *<association>*, the following condition describes the cyclic one-ended |- specification for an association *A*.

$$(A.b_{S|} = |\text{-} \text{ and } A.b_{R|} \neq |\text{-}) \text{ and a link cycle exists including an } A \text{ link}$$

The above condition may or may not result in an ambiguous outcome on an object deletion, explicit link destruction, or link change. Such an outcome will likely result when the link cycle makes it unclear whether or not the single |- binding will be operative. For example, in Fig. 9.7, assume an attempted deletion of object z1. If the y2↔z1 link of A1 is processed first, the |- is inoperative (since the y2↔z1 link of A2 is destroyed on the implicit deletion of y2) and the deletion succeeds; however, if the y2↔z1 link of A2 is processed first, the |- binding is operative and the deletion fails. Which is it? (The reader can verify these outcomes based on Table 2.1 or see Section 10.2.2.1, Case 1 for a more detailed explanation of this example.)

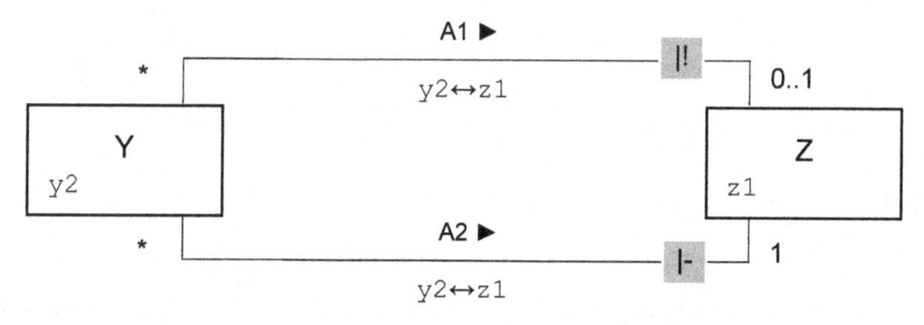

Fig. 9.7 Database model with one-ended |- and database with link cycle z1↔y2, y2↔z1

The second ambiguous specification is a *cyclic non-dependent |' with partly sacrificial siblings*. Again, using the variables defined in the previous sections for describing an <association>, the following condition describes this specification for an association A and an association A', where here A' is an association involving a class related to class S directly by the association A (which may be an intra-class association) or indirectly via A and other associations having a subject class implicit cascade binding of |? or |!.

$$(A.b_{S|} = |! \text{ and } A.m_{SL} \neq A.m_{SU} \text{ and } A'.b_{S|} = \text{nil and } A'.m_{SL} \geq 1 \text{ and } A'.m_{SU} > A'.m_{SL})$$
and a link cycle exists including A and A' links

If the related class described above is directly related to the subject class S of A, i.e., it is the related class R of A, then A' may be A^{-1}.

The above condition may or may not result in an ambiguous outcome on an object deletion, explicit link destruction, or link change. Such an outcome will most likely result when the link cycle makes it unclear which object in the related class, i.e., the subject class of A', will be the "surviving sibling," i.e., the one needed to satisfy $A'.m_{SL}$, and which will be a "sacrificial sibling," i.e, the one not needed to satisfy $A'.m_{SL}$.

For example, in Fig. 9.8, assume an attempted deletion of object x1. If the x1↔y1 link is processed first, the y1 object is implicitly deleted, y1↔z1 is the sacrificial link, the y2 object survives, and the deletion of x1 succeeds; however, if the x1↔y2 link is processed first, the y2 object is implicitly deleted, y2↔z1 is the sacrificial link, the y1 object survives, and again the deletion of x1 succeeds. So, which Y object survives? (Again the reader can verify these outcomes based on Table 2.1 or see Section 10.2.2.2 for a more detailed explanation of this example.)

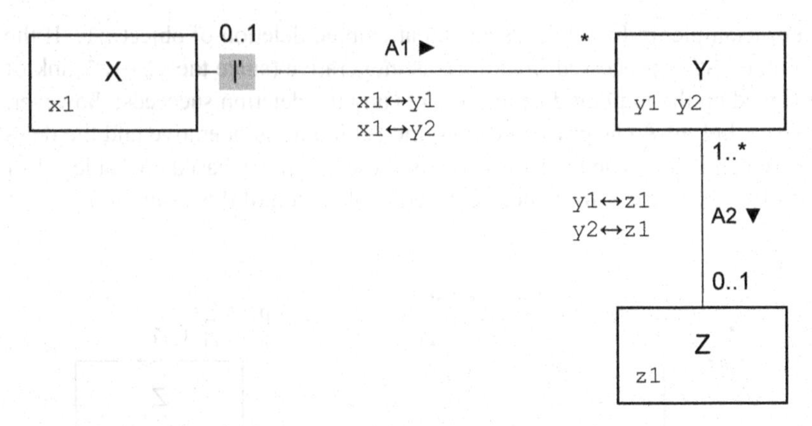

Fig. 9.8 Database model with cyclic non-dependent |' with partly sacrificial siblings and database with link cycle x1↔y1, y1↔z1, z1↔y2, y2↔x1

Modeling tools can check for the possibility of link cycles in the modeled database, detect <*association*> specifications that may result in ambiguous outcomes, and display appropriate warnings. This capability is planned for a future version of the ORN Simulator.

9.4 Associations as Functions

This section provides a functional interpretation of the <*multiplicities*> description of an association and uses this interpretation to show which association types are valid and invalid as intra-class associations. For simplicity, only multiplicities of the form $m..n$ where $m = 0$ or 1 and $n = 1$ or * are considered. The section also provides examples of valid intra-class associations including some symmetric ones.

A <*multiplicities*> description defines two functions f and g corresponding to an association type and its inverse, respectively. Table 9.1 describes the function types of f and g for each possible association type, based on the <*multiplicities*> description. It also indicates whether or not the association type can be an intra-class association. For the purpose of precisely defining the functions f and g, assume that S is the subject class, R is the related class, S' is a subset of S, R' is a subset of R, $P(x)$ is the power set of a set x, and Ø is the empty set.

A *single-valued, total function* f maps S objects to R objects. This is denoted as f: $S \rightarrow R$ where S is the domain and R the codomain. A *single-valued, partial function* f is a function f: $S' \rightarrow R$; a *set-valued (or multi-valued), total function* f is a function f: $S \rightarrow P(R) - Ø$; and a *set-valued, partial function* f is a function f: $S' \rightarrow P(R) - Ø$. A single-valued function f is *injective*, or one-to-one, if and only if no two objects in the domain map to, i.e., are related to, the same object in the codomain. A single-valued function f is *surjective*, or *onto*, if and only if for every object y in the codomain there is an object x in the domain such that $f(x) = y$. A single-valued function f is *bijective* if and only if it is both injective and surjective.

Table 9.1. Associations as Functions

Association	Function Type of f	Function Type of g	Intra-class?
0..1-to-0..1	single-valued, partial, injective	single-valued, partial, injective	Yes
0..1-to-1	single-valued, total, injective	single-valued, partial, bijective	No
1-to-1	single-valued, total, bijective	single-valued, total, bijective	Yes
0..1-to-*	set-valued, partial	single-valued, partial	Yes
0..1-to-1..*	set-valued, total	single-valued, partial, surjective	No
*-to-1	single-valued, total	set-valued, partial	Yes
1-to-1..*	set-valued, total	single-valued, total, surjective	No
-to-	set-valued, partial	set-valued, partial	Yes
-to-1..	set-valued, total	set-valued, partial	Yes
1..*-to-1..*	set-valued, total	set-valued, total	Yes

The definitions of terms given in the previous paragraph are also applicable to the function g with S everywhere replaced by R and R by S.

To show that 0..1-to-1, 0..1-to-1..*, and 1-to-1..* associations are not valid as intra-class associations, let X be a set of objects $\{x_1, x_2, ..., x_n\}$ representing the set of all of the instantiated objects of class X, i.e., the extension of class X.

First, assume that a 0..1-to-1 association exists between the objects of X. Let $f: X \rightarrow X$ be the single-valued, total, injective function and $g: X' \rightarrow X$, where $X' \subseteq X$, be the single-valued, partial, bijective function corresponding to the association and inverse association, respectively. Let $g(x_i)$ be undefined for some $x_i \in X$. Then for all $x \in X, f(x) \neq x_i$, otherwise $g(x_i) = x$ and $g(x_i)$ would be defined. Since f is total, the pigeonhole principle requires that for some $x_l \neq x_i$ and $x_j, x_k, x_l \in X, f(x_j) = f(x_k) = x_l$. But this contradicts the fact that g is single-valued (or that f is injective). Therefore, a 0..1-to-1 association between the objects of X is invalid.

The proof for 0..1-to-1..* is similar. Assume that a 0..1-to-1..* association exists between the objects of X. Let $f: X \rightarrow P(X) - \emptyset$ be the set-valued, total function and $g: X' \rightarrow X$, where $X' \subseteq X$, be the single-valued, partial, surjective function corresponding to the association and inverse association, respectively. Let $g(x_i)$ be undefined for some $x_i \in X$. Then for all $x \in X, x_i \notin f(x)$, otherwise $g(x_i) = x$ and $g(x_i)$ would be defined. Since f is total, the pigeonhole principle requires that for some $x_l \neq x_i$ and $x_j, x_k, x_l \in X, x_l \in f(x_j)$ and $x_l \in f(x_k)$. But this means that $g(x_l) = x_j$ and $g(x_l) = x_k$, which contradicts the fact that g is single-valued. Therefore, a 0..1-to-1..* association between the objects of X is invalid.

Finally, assume that a 1-to-1..* association exists between the objects of X. Let $f: X \rightarrow P(X) - \emptyset$ be the set-valued, total function and $g: X \rightarrow X$ be the single-valued, total, surjective function corresponding to the association and its inverse, respectively. Let the cardinality of $f(x_i)$ be greater than one for some $x_i \in X$. Then, since f is total, there is at least one $x_l \in X$ such that $x_l \in f(x_j)$ and $x_l \in f(x_k)$ for some $x_j, x_k \in X$. But this means that $g(x_l) = x_j$ and $g(x_l) = x_k$, which contradicts the fact that g is single-valued. Therefore, a 1-to-1..* association between the objects of X is invalid.

The validity of the other <*multiplicities*> descriptions as intra-class associations can be shown by simple examples. These are given in Fig. 9.9, where X is assumed to contain objects x_1, x_2, and x_3 and the functions f and g are as defined in Table 9.1. A "--" denotes that the function is undefined for a particular object.

0..1-to-0..1

x	$f(x)$	$g(x)$
x_1	x_2	--
x_2	--	x_1
x_3	--	--

1-to-1

x	$f(x)$	$g(x)$
x_1	x_2	x_3
x_2	x_3	x_1
x_3	x_1	x_2

0..1-to-*

x	$f(x)$	$g(x)$
x_1	$\{x_1, x_3\}$	--
x_2	--	x_1
x_3	--	x_3

*-to-1

x	$f(x)$	$g(x)$
x_1	x_2	--
x_2	x_3	$\{x_1, x_3\}$
x_3	x_2	$\{x_2\}$

-to-

x	$f(x)$	$g(x)$
x_1	$\{x_2, x_3\}$	--
x_2	$\{x_3\}$	$\{x_1\}$
x_3	--	$\{x_1, x_2\}$

-to-1..

x	$f(x)$	$g(x)$
x_1	$\{x_2, x_3\}$	--
x_2	$\{x_3\}$	$\{x_1, x_3\}$
x_3	$\{x_2\}$	$\{x_1, x_2\}$

1..*-to-1..*

x	$f(x)$	$g(x)$
x_1	$\{x_2, x_3\}$	$\{x_3\}$
x_2	$\{x_3\}$	$\{x_1\}$
x_3	$\{x_1\}$	$\{x_1, x_2\}$

Fig. 9.9 Examples of valid intra-class associations

Real associations can be ascribed to each of the examples in Fig. 9.9. For example, an "is a guest of" relationship between persons at a party in which everyone can bring at most one guest is an example on a 0..1-to-0..1 intra-class association. An example of a 1..*-to-1..* intra-class association is an "is reviewed by" association between a class of students where each student must have their work reviewed by at least one other student and each student must review the work of at least one other student.

0..1-to-0..1, 1-to-1, *-to-*, and 1..*-to-1..* intra-class associations may be symmetric. For example, the 0..1-to-0..1 "is a spouse of" association is a *symmetric association*. For such relationships, f and g are the same function.

9.5 Conclusion

This chapter has used ORN to describe inconsistencies and ambiguities that can occur when defining the semantics for an association and for combinations of associations involving common classes. These inconsistencies and ambiguities can occur when using other notations to define association semantics—like the ON DELETE, NOT NULL, and UNIQUE clauses in SQL—or when association semantics are "defined" only by their implementations in application programs, class methods, or SQL constraints and triggers.

The SQL depicted in Fig. 9.10 illustrates this. A foreign key, which must be NOT NULL, in the table Y references table X. An ON DELETE CASCADE is given for this foreign key. Another foreign key, which must be NOT NULL, in table Z references table Y. An ON DELETE NO ACTION is the default for this foreign key. CHECK constraints given for tables X and Y ensure that for both foreign keys every row in the referenced table has at least one matching row in the referencing table. These notations define associations that are likely inconsistent. Deletion of a row in table X causes the deletion of all matching rows (via Y.fk) in table Y, but this likely causes a referential integrity exception since all of the rows in table Y have at least one matching row (via Z.fk) in table Z. So, unless things are somehow "fixed up" by other referential actions, at least one Z.fk value exists that does not reference any row in table Y.

```
...
CREATE TABLE X (
  pk ... PRIMARY KEY,
  ...
  CONSTRAINT C1 CHECK ((SELECT COUNT(*) FROM Y WHERE Y.fk = X.pk) >= 1)
);
CREATE TABLE Y (
  pk ... PRIMARY KEY,
  ...
  fk ... NOT NULL REFERENCES X ON DELETE CASCADE,
  CONSTRAINT C2 CHECK ((SELECT COUNT(*) FROM Z WHERE Z.fk = Y.pk) >= 1)
);
CREATE TABLE Z (
  pk ... PRIMARY KEY,
  ...
  fk ... NOT NULL REFERENCES Y
);
```

Fig. 9.10 Likely inconsistent associations implemented in SQL

With ORN, database designers can craft association semantics during database modeling, thereby permitting problematic semantics to be detected before any coding or even database definition commences. In the above SQL example, the association between the objects represented by tables X and Y is |!<1-to-1..*>, and that between Y and Z is <1-to-1..*>, **although it is very hard to tell**. When using an ORN-extended modeling tool, like the ORN Simulator, these <*association*>s properly result in a warning of inconsistent associations.

Part III

Adding ORN to a DBMS

Chapter 10

A Conceptual Implementation of ORN

Exploring Semantic Circularity and Ambiguity

This chapter presents algorithms that implement ORN semantics. The algorithms are given at a conceptual level in terms of classes, objects, associations, and links. They provide context for and more detail on the system actions given for each binding in Table 2.1, thus providing a very precise algorithmic description of ORN semantics.

Moreover, the algorithms outline the code that translation tools must generate to implement ORN in a database system. In OR+, discussed in Chapter 7, this code is implemented as methods on abstract, persistent classes that are defined to an ODBMS. In ORN Additive, discussed in Chapter 6, this code is implemented as stored procedures and triggers on ORN-related tables that are defined to a RDBMS. Chapter 12 recasts the algorithms given here in terms compatible for their implementation in a standard ODBMS (Cattel 2003). The algorithms can also be recast for implementation within an RDBMS to support the integration of ORN into SQL as discussed in Chapter 11. This task is left to RDBMS developers and would provide a much more efficient implementation than that provided by ORN Additive.

Besides presenting algorithms that implement ORN, this chapter shows that these algorithms terminate despite processing cyclically linked data and that their results are deterministic, with two exceptions. The former characteristic ensures that ORN semantics are non-circular and the later ensures they are mostly unambiguous.

We examine both characteristics in the context of link cycles within a database. A *link cycle* occurs when an object is related to itself indirectly—i.e., $x_0 \leftrightarrow x_1$, $x_1 \leftrightarrow x_2$, ..., $x_n \leftrightarrow x_0$ where $n \geq 1$. The potential problems posed by such cycles to ORN semantics—infinite and alternative processing paths, resulting in circularity and ambiguity—are inherent in any scheme that defines association semantics recursively, as does ORN.

This chapter significantly improves on the algorithmic description of ORN and related discussions on link cycles, circularity, and ambiguity first given in Ehlmann et al. (2002).[1] It is organized as follows. Section 1 presents and explains the high-level algorithms. Section 2 explores their operation in the context of link cycles, providing examples of ORN-defined associations and sample databases that demonstrate algorithm termination and two kinds of problematic ORN specifications that can result in nondeterministic results. Possible solutions are discussed, and a theorem with proof is given that ORN semantics are unambiguous in the absence of the two problematic specifications. Section 3 concludes with some summary remarks.

---

[1] Portions reprinted, with permission, from "Specifying and enforcing association semantics via ORN in the presence of association cycles," IEEE Trans on Knowl and Data Eng 14, 6:1249-1257. © 2002 IEEE.

B.K. Ehlmann, *Object Relationship Notation (ORN) for Database Applications*,
Advances in Database Systems 39, DOI 10.1007/978-0-387-09554-7_10,

10.1 Algorithms

The implementation of ORN semantics is described by the algorithms that create objects and association links, delete objects, and destroy and change association links in a database. These operations become complex object operations in the context of ORN. In this section, we describe these algorithms.

The conceptual nature of the algorithms makes them independent of a particular implementation, object or relational. A link between objects x and y, mathematically an ordered pair (x, y), is represented in the algorithms as $x \leftrightarrow y$ where x is the subject object and y is the related object. The type of a link is the association of which it is an instance. The objects of a link, along with its type, make it unique. Every association A has an inverse association, A^{-1}, where the roles of subject and related class are reversed. If $x \leftrightarrow y$ is a link of type A, $y \leftrightarrow x$ is a link of type A^{-1}.

This section indicates how the conceptual algorithms can translate into an actual relational or object database implementation. In an object database, an association is represented by an object-based attribute, e.g., customer in the class Order (Fig. 1.18), and the inverse association by an inverse object-based attribute, e.g., orders in class Customer. In a relational database, an association is represented by a foreign key attribute, e.g., custNo in table Order (see Fig. 1.8), and the inverse association is represented only implicitly by the rows in the referencing table having matching foreign key values to the primary key value in each row of the referenced table.

In addition to providing a conceptual view of the actual implementation of ORN in OR+, which is discussed in Chapter 7, the algorithms provide a simplified, abstract view of this implementation. For instance, they do not show the details of handling all types of errors, e.g., improper transaction operations. They also do not show all of the details needed to properly handle association inheritance.

The algorithms are given in a pseudocode where compound statements within control structures are indicated by indentation. Also, the input, output, and input/output nature of parameters is implicit from the comments describing the algorithm. Comments are prefaced by a //. Finally, the function type(x) is used, which when x is an object, returns the most specific class in any class hierarchy for which x is an instance, and when x is a link, returns the association for which x is an instance.

10.1.1 Algorithm CreateObject

When an ORN-relatable object is created (or instantiated) in a database—e.g., via an object creation operator, like new in C++ or Java or an INSERT statement in SQL—the implementation of ORN must ensure that the complex object is properly constructed. This means that all lower bound multiplicities for any associated classes are satisfied before the transaction containing the object creation can commit.

Fig. 10.1 gives the algorithm for creating an object of a particular type. In addition to creating the primitive object, described here via a new operation, the algorithm adds an element to the set *t.LbChecks* for each association involving the object that has a lower bound multiplicity constraint for the related class. *LbChecks* is associated with the current application-defined transaction, whose existence is assumed by CreateObject and referenced by *t*. Creation and modification of relatable objects must take place within such a transaction. When it commits, checks are made on each existing object and association referenced in *LbChecks* to ensure that lower bound multiplicities for related classes are not violated.

```
Algorithm CreateObject(o: Object, C: Class, t: Transaction, d: Database)
// Create an object o of type C in database d. (Assumes prior Begin on transaction t.)
 o = new C;
 for each association A where C is the subject class do
   rLb = lower bound multiplicity for related class of A;
   if rLb > 0 then add (o, A) to t.LbChecks;
 exit(successful);
```

Fig. 10.1 Algorithm for creating an object

```
Algorithm Begin(t: Transaction, d: Database)
// Begin transaction t on database d.
 Set t.LbChecks and t.Deletes to empty.
 Perform other begin transaction functions;
 if exception then exit(exception);
 exit(successful);

Algorithm Abort(t: Transaction, d: Database)
// Abort transaction t on database d.
 Perform abort transaction functions;
 if exception then exit(exception);
 exit(successful);

Algorithm Commit(t: Transaction, d: Database)
// Commit transaction t on database d.
 for each object, association (x, A) in t.LbChecks do
   if x is not in t.Deletes or the Deletes of any ancestor transaction then
     rLb = lower bound multiplicity for related class of A
     if rLb is violated then exit(exception);
 Perform other commit transaction functions;
 if exception then exit(exception);
 if t is a nested transaction then Add objects in t.Deletes to Deletes of parent transaction;
 exit(successful);
```

Fig. 10.2 Algorithms for beginning, aborting, and committing a transaction

10.1.2 Transaction operations

Algorithms for the transaction commit operation as well as those for transaction begin and transaction abort are given in Figure 10.2. These operations perform the standard database functions required for beginning, aborting, and committing a transaction. They must, however, be extended for ORN to perform actions related to checking lower bound multiplicities and tracking object deletions. They also must support nested transactions.

The Begin algorithm initializes two object sets, *Deletes* and *LbChecks*, to empty. These sets are associated with every transaction. *Deletes* is the set of all objects marked for deletion so far by the transaction, including objects marked for deletion by any committed nested transactions. *LbChecks*, discussed previously, is the set of all (object, association) pairs where the object is an instance of the subject class in the given association and where for this object the lower bound multiplicity for the related class must be checked at transaction commit, provided the object has not been previously marked for deletion. The Commit algorithm shows how the sets *Deletes* and *LbChecks* are used to do the appropriate lower bound constraint checking.

10.1.3 Supporting pseudocode for complex object operations

This subsection presents pseudocode that supports the algorithms for the complex object operations CreateLink, DeleteObject, DestroyLink, and ChangeLink. This pseudocode provides a wrapper algorithm, which invokes the complex object operation, and a sub-algorithm, which is invoked by three of these operations.

The algorithms for the complex object operations are invoked within a system-supplied nested transaction, providing a wrapper for each such algorithm. The nested transaction, given in Fig. 10.3, results when a complex object operation is specified within an application-defined transaction. A translation of the specific syntax for this operation must provide the appropriate arguments to the corresponding CreateLink, DeleteObject, DestroyLink, or ChangeLink algorithm as defined in a subsequent subsection. The nested transaction ensures that these operations are atomic.

```
Begin(t, d);
Invoke the CreateLink, DeleteObject, DestroyLink, or ChangeLink algorithm corresponding to the
  specified complex object operation;
if exception then
  Abort(t, d);
  raise exception
else
  Commit(t, d)
  if exception then
    Abort(t, d);
    raise exception
```

Fig. 10.3 Invocation of a complex object operation in a nested transaction

The algorithms for DeleteObject, DestroyLink, and ChangeLink invoke the sub-algorithm EnforceBinding, given in Fig. 10.4. As its name implies, this sub-algorithm enforces an ORN binding. The reader should compare its pseudocode to the system actions described for each binding in Table 2.1.

```
Algorithm EnforceBinding(b: Binding, A: Association, sO: Object, rO: Object,
                          t: Transaction, d: Database)
// Enforce binding b for subject class in association A upon destruction of link between subject
// object sO and related object rO, i.e., sO↔rO.
  sLb = lower bound multiplicity for the subject class of A;
  case b
    nil:  if sLb is violated then insert (rO, A^-1) into t.LbChecks;
    "-":  exit(exception);
    "?":  if sLb is violated then
            DeleteObject(rO, t, d);
            if exception then exit(exception);
    "!":  DeleteObject(rO, t, d);
          if exception then exit(exception);
    "'":  if sLb is violated then insert (rO, A^-1) into t.LbChecks;
          Begin(nT, d);
          DeleteObject(rO, nT, d);
          if exception then Abort nT
          else Commit(nT, d);
               if exception then Abort nT;
  end case
  exit(successful);
```

Fig. 10.4 Algorithm for enforcing ORN bindings

When the lower bound multiplicity is violated on a default, i.e., nil, or ' binding, an exception does not immediately result; rather, the related object and applicable association are inserted into the set *t.LbChecks*, deferring any possible exception until the end of the operation, i.e., the nested transaction commit. These actions effectively implement the semantic specified by footnote 2 in Table 2.1 and will be important to remember in the next section.

The ' binding requires yet another nested transaction to appropriately implement its semantics. If an exception occurs within this transaction, it is simply ignored and the transaction is aborted; however, a deferred lower bound check on the commit of the parent, nested transaction can still result in an exception.

The cascade binding cases in EnforceBinding can result in a recursive invocation of DeleteObject. This recursion is the focus of much discussion in the next section.

10.1.4 Algorithm **CreateLink**

The CreateLink algorithm is given in Fig. 10.5. "Create link *x* ↔ *y* of type *A*" conceptually implements an association instance. At a logical database level, this usually involves creating the necessary references between the two objects. In an object database, this means setting or inserting appropriate references into the object-based attributes of the subject and related objects. In a relational database, this means setting the foreign key to an appropriate primary key value.

```
Algorithm CreateLink(A: Association, x: Object, y: Object, t: Transaction, d: Database)
// Create a link of type A between object x and object y in database d. (Assumes prior Begin on
// transaction t.)
   sC = type(x); // subject class
   rC = type(y); // related class
   sUb = upper bound multiplicity for sC of A;
   rUb = upper bound multiplicity for rC of A;
   if sUb or rUb is violated by creation of x↔y then exit(exception);
   Create link x↔y of type A;
   exit(successful);
```

Fig. 10.5 Algorithm for creating an association link

10.1.5 Algorithm DeleteObject

The DeleteObject algorithm, given in Fig. 10.6, deletes an object x and appropriate related objects as defined by ORN. Every association involving the object x is accessed by the outer **for each** loop. In the inner **for each**, an existing link is implicitly destroyed before any implicit related object deletion is attempted in EnforceBinding. Thus, the destroyed link is not considered when determining whether or not an implicit deletion of the related object is possible—that is, it is no longer an **existing** link.

```
Algorithm DeleteObject(x: Object, t: Transaction, d: Database)
// Delete complex object x in database d. (Assumes prior Begin on transaction t.)
   if x in t.Deletes or the Deletes of any ancestor transaction then exit(successful);
   Insert x into t.Deletes;
   C = type(x);
   for each association A defined where class C is the subject class (in the order defined) do
      impB = implicit destructibility binding for class C in A;
      for each link x↔rO of type A where x is the subject object and rO is a related object do
         Destroy link x↔rO;
         EnforceBinding(impB, A, x, rO, t, d);
         if exception then exit(exception);
      end for
   end for
   Delete primitive object x;
   exit(successful);
```

Fig. 10.6 Algorithm for deleting an object

"Destroy link x↔rO" in an object database, of course, involves destroying the reference from object x to object rO and the reference from rO to x. In a relational database it involves setting a foreign key to NULL (or at least simulating it).

The invocation of EnforceBinding results in recursion when this sub-algorithm invokes DeleteObject. Such recursion implements a depth-first traversal through *d* to implicitly delete all appropriately related objects as defined by ORN.

10.1.6 Algorithm DestroyLink

The DestroyLink algorithm, given in Fig. 10.7, destroys a link and via EnforceBinding may cascade this destruction to appropriately related objects as defined by ORN. The algorithm invokes EnforceBinding twice to enforce the bindings at each end of the association for the link, treating in turn each object in the link as the subject object.

```
Algorithm DestroyLink(x↔y: Link, t: Transaction, d: Database)
// Explicitly destroy link x↔y in database d. (Assumes prior Begin on transaction t.)
  A = type(x↔y);
  Destroy link x↔y;
// Enforce explicit binding from the perspective of the subject class of A.
  sO = x;  rO = y;  C = type(sO);
  expB = explicit destructibility binding for class C in A;
  EnforceBinding(expB, A, sO, rO, t, d);
  if exception then exit(exception);
  // Enforce explicit binding from the perspective of the related class of A.
  sO = y;  rO = x;  C = type(sO);
  expB = explicit destructibility binding for class C in A;
  EnforceBinding(expB, A^-1, sO, rO, t, d);
  if exception then exit(exception);
  exit(successful);
```

Fig. 10.7 Algorithm for destroying an association link

10.1.7 Algorithm ChangeLink

The ChangeLink algorithm, given in Fig. 10.8, is essentially an explicit destruction of a link for an association *A* between a subject class object x and a related class object *y1*, followed by the creation of a new *A* link between x and a different related class object *y2*. A major difference, however, is that the explicit destructibility binding, *expB*, for the related class is not enforced. Lower bound multiplicities for this class will never be violated since one related class object is simply being replaced by another. Also, a check is made to ensure that objects x and *y2* still remain before the x↔*y2* link is created since the enforcement of the explicit destructibility binding for the subject class could have implicitly deleted one or both of these objects.

```
Algorithm ChangeLink(x↔y1: Link, y2: Object, t: Transaction, d: Database)
// Change link x↔y1 in database d replacing the related object y1 with y2. (Assumes prior Begin
// on transaction t.)
  A = type(x↔y1); C = type(x);
  // Destroy the link between x and y.
  Destroy link x↔y1;
  expB = explicit destructibility binding for class C in A;
  EnforceBinding(expB, A, x, y1, t, d);
  if exception then exit(exception);
  // Create a link between x and y2.
  if x is in t.Deletes or y2 is in t.Deletes then exit(exception);
  sUb = upper bound multiplicity for C in A;
  if sUb is violated by creation of x↔y2 then exit(exception);
  Create link x↔y2 of type A;
  exit(successful);
```

Fig. 10.8 Algorithm for changing an association link

In an object database, a link is changed when a reference in a single-valued, object based attribute is changed from one non-NULL value to another or is replaced in a multi-valued, object-based attribute with another. In a relational database, a link is changed when a value in a foreign key is changed from one non-NULL value to another.

10.2 Link Cycles

Others have studied the problems posed by "link cycles" within relational databases and SQL (Date 1990, Date and Darwen 1994, Howowitz 1992), although in these studies such cycles were called *referential cycles*, and the concern was not in maintaining multiplicities as with ORN, but rather in maintaining referential integrity. Some of this previous work, however, is germane to our exploration of link cycles in this section where we deal at the entity or object level.

Actually, what was defined as a referential cycle in this previous, relational database work would be called an *association cycle* in a conceptual data model. An association cycle occurs in a model when a class is related to itself directly via an intraclass association or indirectly via associations with other classes. An association cycle in a model can result in a link cycles in a database as we shall see; however, a link cycle can occur in a database even without the presence of association cycles in the database model or DDL, as we shall also see.

To investigate the problems caused by link cycles, we will study some simple examples of such cycles.

10.2.1 Circularity

Fig. 10.9 depicts one example. An association cycle is present in the model because objects of class Y can be related to the objects of class Z via the A1 association and the objects of class Z can be related to the objects of class Y via the A2 association. There are just two objects in the database, y1 and z1, and two links, y1↔z1 of association A1 and y1↔z1 of A2. The association cycle in a database model allows link cycles to develop in the database, here the cycle is z1↔y1, y1↔z1.

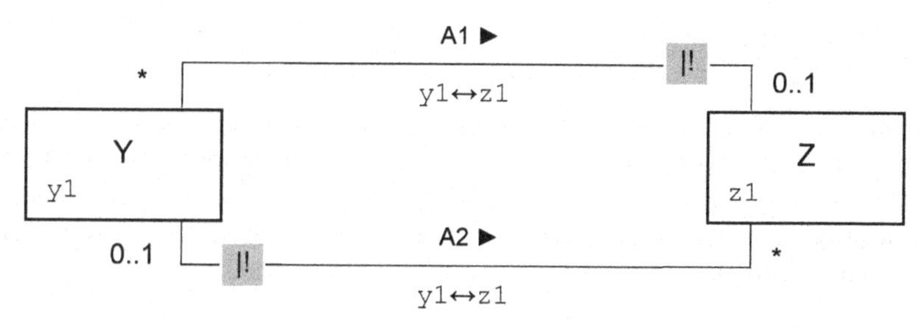

Fig. 10.9 Link cycle z1↔y1, y1↔z1

In this and subsequent examples, we assume ORN semantics as implemented by the algorithms given in the previous section and examine what happens when an attempt is made to delete an object, here z1, in the context of a link cycle.

In Fig. 10.9, there are two possible scenarios.

- If A1 (or more precisely its inverse $A1^{-1}$) is processed first, trying to delete z1 causes an implicit destruction of the y1↔z1 link of A1 (or the z1↔y1 link of $A1^{-1}$) and an implicit delete on y1. This is based on the |! binding for class Z in the A1 association. The implicit delete of y1 will result in the implicit destruction of the y1↔z1 link of A2 and an implicit delete of z1 (based on the |! binding for Y in the A2 association). This will be successful since z1 has previously been marked for deletion—i.e., the recursive call to DeleteObject, in Fig. 10.6, exits successfully since *x* (here z1) is in *t.Deletes*. Now, when A2 (or more precisely $A2^{-1}$) is processed for z1 to see if links exist that require implicit destruction, none is found. Thus the delete of z1 is successful.
- If A2 is processed first, trying to delete z1 causes an implicit destruction of the y1↔z1 link of A2 (based on the implicit default binding and * multiplicity for Z in the A2 association). Next, A1 is processed, which causes an implicit destruction of the y1↔z1 link of A1 and an implicit delete on y1, which will be successful. Thus the delete of z1 is again successful.

One problem with link cycles is that the recursion inherent in the semantics of ORN, and often in those of similar declarative schemes, is circular (and thus algorithmically infinite) unless there is some means to detect a link cycle. As the first scenario above shows, the DeleteObject algorithm detects a cycle and terminates recursion by means of the set *Deletes*. Objects are marked for deletion by placing them into this set. Then, recursive propagation of implicit deletes is terminated when an object to be deleted is found in *t.Deletes* or the *Deletes* of any ancestor transaction, i.e., when a link cycle is detected.

Note that in deleting z1 via the DeleteObject algorithm, as described above, the order in which the associations were processed did not matter. Unfortunately, this is not always the case.

10.2.2 Ambiguities

Two kinds of ORN specifications can potentially cause the algorithms implementing ORN to have nondeterministic, i.e., unpredictable or ambiguous, results. One involves the |- binding and the other the |' binding when these bindings are specified for associations having links within link cycles. These specifications were briefly described in Section 9.3. Here, we analysis them in more detail.

10.2.2.1 The "cyclic one-ended |-"

Fig. 10.10 depicts another link cycle. It is a simplified, non-relational version of an example first given in Date (1990). We again examine what happens when an attempt is made to delete z1. In Fig. 10.10, *b* indicates a possible implicit destructibility binding. We look at two cases.

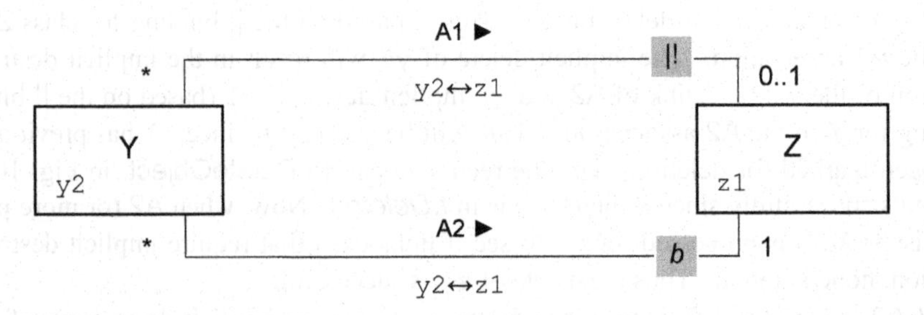

Fig. 10.10 Link cycle z1↔y2, y2↔z1

For Case 1, assume *b* is replaced by a |- binding. There are two scenarios.

- If A1 is processed first, trying to delete z1 causes an implicit destruction of the y2↔z1 link of A1 and an implicit delete on y2. This will be successful and result in the implicit destruction of the y2↔z1 link of A2. Now, when A2 is processed for z1 to see if links exist that require implicit destruction, none is found. Thus the delete of z1 is successful.

- If A2 is processed first, trying to delete z1 will be unsuccessful because the |-binding prevents the destruction of the y2↔z1 link of A2. The complex object operation will be rolled back.

Note that changing the multiplicity for Z in the A2 association from 1 to 0..1, would not change the above scenarios. Also, if the implicit binding for Y in association A2 was |- instead of default, the delete of z1 would always be unsuccessful.

For Case 2, assume *b* in Fig. 10.10 is replaced by an implicit default binding.

- If A1 is processed first, trying to delete z1 again causes an implicit destruction of the y2↔z1 link of A1 and an implicit delete on y2. This will again be successful resulting in the implicit destruction of the y2↔z1 link of A2. Now, when A2 is processed to see if any links require implicit destruction, again none is found. Thus the delete of z1 is successful.
- If A2 is processed first, trying to delete z1 causes an implicit destruction of the y2↔z1 link of A2. This would seem to result in a multiplicity violation of the lower bound 1. However, no action is taken on this violation at this time, instead the check on this constraint is deferred to the end of the complex object operation, i.e., to the commit of its encompassing nested transaction. (In case nil of the EnforceBinding algorithm, Fig. 10.4, a reference to the related object y2 and association A2, i.e., (y2, A2), is inserted into *t.LbChecks*.) Next, A1 is processed, which causes an implicit destruction of the y2↔z1 link of A1 and an implicit delete on y2, which will be successful. At commit of the complex object operation, no constraint violation for (y2, A2) is found (or even checked for) since y2 does not exist (i.e., the Commit algorithm in Fig. 10.2 detects that y2 is in *t.Deletes*). Thus, the delete of z1 is successful.

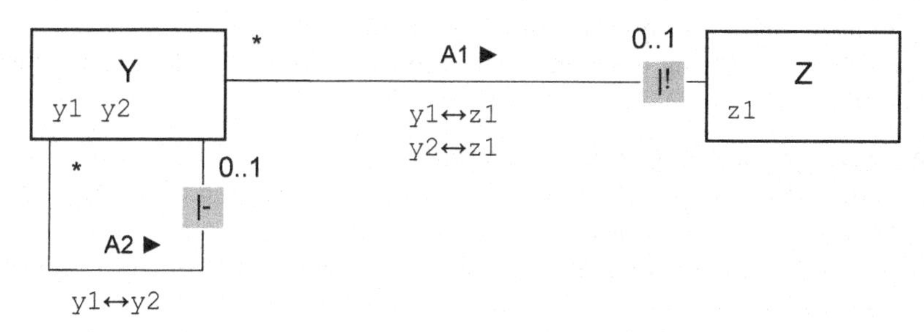

Fig. 10.11 Link cycle z1↔y1, y1↔y2, y2↔z1

Fig. 10.11 is an example of a link cycle involving links of the same association. Again, we examine what happens when an attempt is made to delete z1.

- If the y1↔z1 link of A1 is processed first, trying to delete z1 causes an implicit destruction of this link and an implicit delete on y1, which will be successful and result in the implicit destruction of the y1↔y2 link of A2. Now, when the y2↔z1 link of A1 is processed, this link is implicitly destroyed, and an attempt is made to delete y2, which will succeed since the y1↔y2 link no longer exists. Thus the delete of z1 is successful.

- If the y2↔z1 link of A1 is processed first, trying to delete z1 causes an implicit destruction of this link and an implicit delete on y2, which will be unsuccessful because the |- binding prevents the destruction of the y2↔y1 link of $A2^{-1}$. The complex object operation will be rolled back.

Again, note that if the binding on the * end of association A2 was |-, the delete of z1 would always be unsuccessful.

Besides the potential circularity problem, another problem with link cycles is evident from the above examples. They can cause the outcome of a complex object operation to be dependent on the order in which associations and links are processed. This can occur because cycles provide two alternate processing paths from one object to another and those paths can have different semantics. Fig. 10.10, case 1 shows that outcomes can be dependent on the order in which different associations are processed, and Fig. 10.11 shows that outcomes can be dependent on the order in which the links of a single association are processed. When processing order is unspecified or indeterminate—as it is in relational database definitions and formal mathematical notations, both involving iterations over (unordered) sets—undesirable anomalies can occur when within an implementation an ordering must be selected (Date 1990).

There are many ways to avoid this unpredictability. The following list borrows from Date (1990).

1. Ideally, we could redesign the language or notation so that there is no loss in functionality and processing order does not matter.
2. We could somehow allow the user to specify the processing order when it matters.
3. The system could try all possible processing orders at run-time and always fail if any of them fail (or always succeed if any succeed).
4. Cases where processing order may matter could be detected at definition time and be disallowed.

The reader can probably discern the relative merits of each of these solutions. In the evolution of ORN we have used solution 1, and a database developer can currently employ 2 to some extent in OR+.

The |- binding can cause processing order dependencies only when it is given for just one end of an association having links within a link cycle. This is evident in the previous scenarios and is formally proven in Section 10.2.3. The |- binding is similar to the RESTRICT referential integrity rule in SQL and suffers the same problem (Date and Darwen 1994, Horowitz 1992). Unlike the RESTRICT, however, the |- binding can be protected from a "rear attack" by specifying this same binding for both ends of an association. In Figs. 10.10 and 10.11, when the |- binding is given for both ends of the A2 association, dependencies on the order of processing are eliminated, and the delete of z1 is always unsuccessful.

Use of the |- on only one end of an association when links of the association can form a link cycle, i.e., a *cyclic one-ended* |-, is often desirable and harmless, which is why it is not simply disallowed. In Fig. 10.12, a one-ended |- is used where no unpredictability results even though the intra-class association results in an association cycle in the model, which results in a link cycle in the database. Child c1 is supported by employee e1, who is married to e2, who also supports c1 (c1↔e1, e1↔e2, e2↔c1). When a cyclic one-ended |- results in unpredictability, solution 4 above could be adopted to disallow it, but this was not done in OR+ since link cycles are not inevitable when association cycles exist in the database model and a warning can be issued.

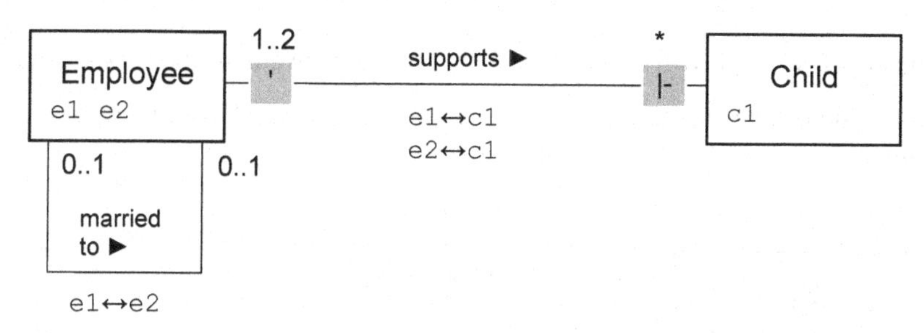

Fig. 10.12 Link cycle c1↔e1, e1↔e2, e2↔c1

Solution 2 can sometimes be employed when a cyclic one-ended |- results in processing order dependencies, which is hopefully rare. In OR+, a user can indirectly specify and predict the ordering in which associations and links are to be processed. Associations for an object are processed in the order in which their associated object-based attributes are declared in the object's class, and links for an association are processed in the order in which an iterator over a multi-valued, object-based attribute (or collection) returns references to the related objects. To control this ordering the user can use an ordered collection, e.g., a List versus a Set, to implement the association. Note that such "link processing order control" can only be used in a relational database at an internal level.

The previous solution, however, is not highly desirable; hence, cases of processing order dependencies should be avoided. Sometimes they can be avoided by replacing a cyclic one-ended |- binding with an implicit default binding and 1 multiplicity. In some respects, this combination is similar to the NO ACTION referential integrity rule in SQL (Date and Darwen 1994) and, as seen in the Case 2 scenario of Fig. 10.10, avoids any order dependency problems.

10.2.2.2 The "cyclic non-dependent |' with partly sacrificial siblings"

As shown in Chapters 4 and 5, the |' (implicit tentative cascade) binding provides powerful semantics that are useful in describing a number of association types. These semantics, however, come with a small price. It is possible (though rare) that ssociations are defined where the precise results of an object deletion can be ambiguous. The necessary conditions for this ambiguity are rather esoteric. They are:

- A class in an association has an implicit ' binding and a multiplicity *m..n* where $m \neq n$, allowing related objects to be non-dependent.
- A related class has an association with an implicit default binding and a multiplicity *m..n* where $m \geq 1$ and $n > m$ for the subject class. This related class is directly related by the association having the |' binding (which may be an intra-class association) or it is indirectly related via this association and other associations having a subject class implicit cascade binding unlike the ' binding being described.
- a link cycle exists involving objects related by the above described associations.

This situation can be characterized as a *cyclic non-dependent |' with partly sacrificial siblings.*

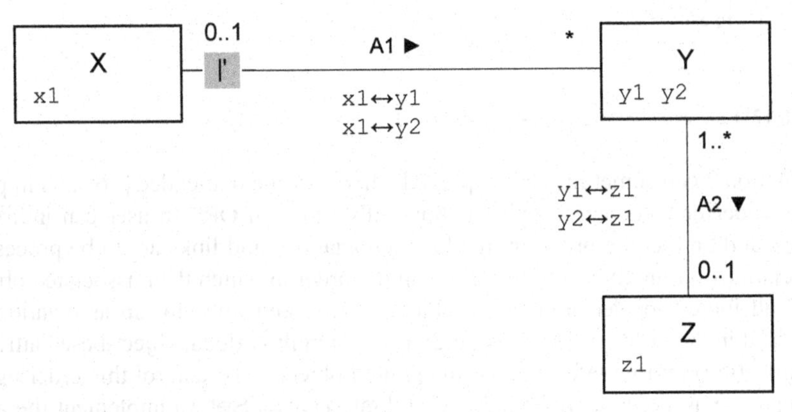

Fig. 10.13 Initial database with link cycle x1↔y1, y1↔z1, z1↔y2, y2↔x1

The model shown in Fig. 10.13 shows such a non-dependent |' binding and the database exemplifies a link cycle involving this binding and some partly sacrificial siblings. The 0..1 multiplicity for class X in association A1 makes the |' binding "non-dependent." (y1 and y2 are not dependent on x1). The "siblings" here are y1 and y2, both being "children" of z1. The fact that there are two of them related to z1 when the multiplicity for Y in association A2 is 1..* makes them "partly sacrificial," i.e., one of them can be deleted. We will examine what happens when an attempt is made to delete x1. There are two possible scenarios.

- If the x1↔y1 link of A1 is processed first, trying to delete x1 causes an implicit destruction of this link and an attempted implicit delete on y1, which will be successful and result in the implicit destruction of the y1↔z1 link of A2. This latter link is destructible since the implicit binding for Y in the A2 association is default and the 1..* multiplicity is not violated. Now, when the x1↔y2 link of A1 is processed, this link is implicitly destroyed, and an attempt is made to implicitly delete y2. This, however, does not succeed since the lower bound multiplicity of 1 (for Y in A2) is violated. (This is detected by the Commit algorithm, Fig. 10.2, which was called in case "!" of the EnforceBinding algorithm, Fig. 10.4, which was called by the DeleteObject algorithm, Fig. 10.6, which was called to delete x1.) Basically, z1 must be related to at least one Y object. Thus the attempted implicit deletion of y2 is aborted (in EnforceBinding), the deletion of x1 is successful (since deletion of y2 is not required), and the database shown in Fig. 10.14 results.

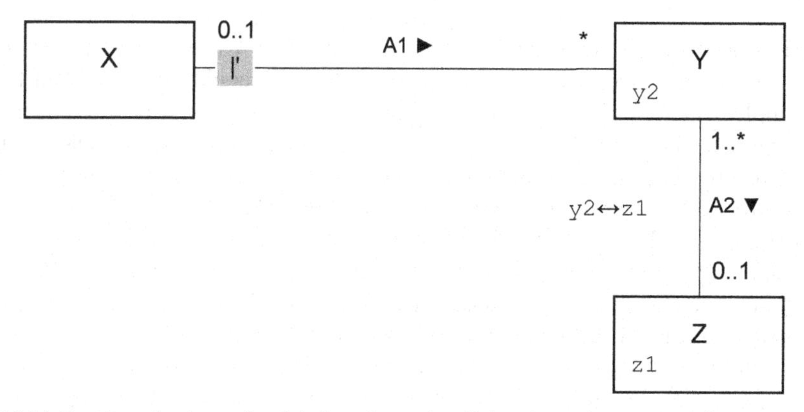

Fig. 10.14 Resultant database after deletion of x1 when link x1↔y1 is processed first

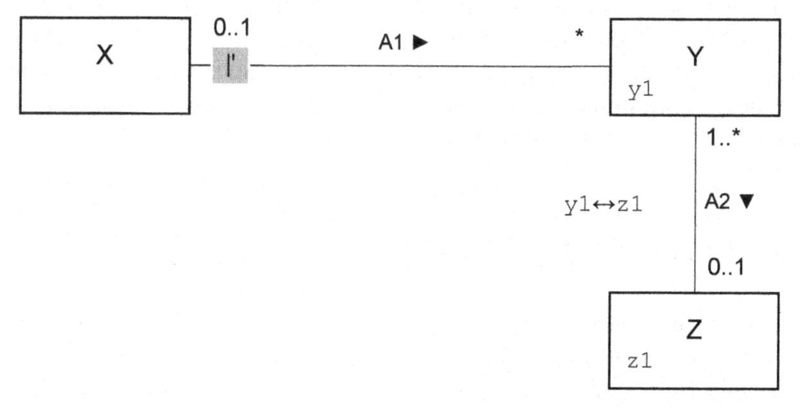

Fig. 10.15 Resultant database after deletion of x1 when link x1↔y2 is processed first

- If the x1↔y2 link of A1 is processed first, trying to delete x1 causes an implicit destruction of this link and an attempted implicit delete on y2, which will be successful and result in the implicit destruction of the y2↔z1 link of A2. Now, when the x1↔y1 link of A1 is processed, this link is implicitly destroyed, and an attempt is made to implicitly delete y1. This, however, does not succeed since the lower bound multiplicity of 1 is violated, i.e., z1 must be related to at least one Y object. Thus the attempted implicit deletion of y1 is aborted, the deletion of x1 is successful, and the database shown in Fig. 10.15 results.

As can be seen by the above scenarios, the |' binding ensures that all related objects **that are not required** are deleted, and when the requirement is that a minimum number of related objects exist for some object, precisely which of the related objects remain is immaterial. So, in some sense, the result of an x1 object deletion is unambiguous. One Y object along with its link to a Z object will always remain. In a strict sense, however, the result is ambiguous since exactly which Y object remains is unclear.

If such unpredictability is unacceptable, solutions 2 or 4 as explained in the previous subsection could be used. An association like A1 could be disallowed by an ORN implementation. Note, however, a link cycle occurs here when no association cycle exists in the model! Alternatively, in an implementation, like OR+, a user could indirectly specify the ordering in which links are to be processed by controlling the order in which references are placed into ordered collections.

Likely, the need for associations using the |' binding in a context similar to that just examined is rare. Note, for instance, that if the multiplicity in Fig. 10.13 for class X in association A1 was 1 instead of 0..1 (resulting in a dependency), no ambiguity would result on deletion of an X object. The deletion of x1 in Fig. 10.13 would always fail since at the commit of this complex object operation, either y1 or y2 would be left without a related X object as seen in Figs. 10.14 or 10.15, respectively (i.e., in the Commit invoked in Fig. 10.3 a lower bound multiplicity violation would result in an exception). In general, an X object could only be deleted if it was not related to any Y objects whose association with a Z object might be required. Also, note that if the |' binding was changed to a |! binding in Fig. 10.13, no ambiguity would result since the deletion of x1 would always fail since z1 would be left without a Y object (i.e., in the Commit invoked in Fig. 10.3 a lower bound multiplicity violation would result in an exception).

10.2.3 The Theorem for ORN Semantic Clarity and its proof

Here we state and prove a theorem that ORN semantics are unambiguous in the absence of the two problematic bindings discussed in the previous subsection. The proof for this *Theorem for ORN Semantic Clarity* is quite lengthy but includes some explanation, using examples given in the previous subsection.

Theorem: If no cyclic one-ended |- binding and no cyclic non-dependent |' binding with partly sacrificial siblings are given for an association, then the semantics of ORN are unambiguous—that is, the outcome of a complex object operation under ORN semantics is independent of the order in which links are processed.

Proof: A complex object operation is a link creation or an explicit object deletion, link destruction, or link change. As can be seen in Fig. 10.5, the LinkCreate algorithm involves no recursion, and while it may create link cycles, it is unaffected by them. It processes only one link and thus results are independent of the order in which links are processed.

We next consider explicit object deletion.

If the object being deleted and all objects linked to it directly or indirectly are not part of any link cycle, then there is only one processing path to any related object or link and thus only one possible outcome.

If, however, the object being deleted or any object linked to it directly or indirectly is part of one or more link cycles, then there can be multiple processing paths to related objects and links. We must show that the result of an object deletion will be unaffected by the order in which links are processed. We do this by showing that the result of executing the DeleteObject algorithm, invoked in a nested transaction *t*, as shown in Fig. 10.3, to delete an object *x* in database *d*, is unaffected by the order in which the links of *x* or any related object are processed.

This result, denoted by *R*, is determined by whether or not exit is with exception, and if not, the elements that are contained in three sets at the time of commit of any transaction *t*, where *t* is the transaction wrapping the complex object operation (see Fig. 10.3) or any nested transaction within this transaction. The three sets are:

- the set *t.Destroys* of links that have been destroyed, which are all the links destroyed in transaction *t* and all ancestor transactions
- the set *t.Deletes'* of objects that are or will be deleted, which is the union of *t.Deletes* and the set *Deletes* for all ancestor transactions
- the set *t.LbChecks'* of object-association pairs defining the lower bound multiplicity checks that must be done at the commit of *t*, which is $\{(sO, A) \mid (sO, A) \in$ *t.***LbChecks** $\wedge\, sO \notin$ *t.Deletes'*$\}$

Note that the sets *t.Destroys*, *t.Deletes'*, and *t.LbChecks'* determine the results of the lower bound multiplicity checks, so that these results are not an issue.

Let *o* be *x* or any object that is related to *x* directly or indirectly. Assume that prior to the invocation of DeleteObject(*x*, *t*, *d*), *o* has *n* links to related objects, $o \leftrightarrow o_1, o \leftrightarrow o_2, \ldots, o \leftrightarrow o_n$. The links may involve one or more association types, the *n* related objects may not all be unique and may in fact be *o*, and *o* may be part of one or more link cycles.

If *o* is being explicitly deleted, then prior to DeleteObject(*o*, *t*, *d*), none of *o*'s links have been implicitly destroyed. If, however, *o* is being implicitly deleted, then prior to DeleteObject(*o*, *t*, *d*) one of its links, i.e., an *entry link*, has already been destroyed—e.g., the `y2↔z1` link of A1 for object `y2` in Fig. 10.10, Case 1, when A1 is processed first.

Furthermore, if *o* is part of one or more link cycles, then before a link can be processed by DeleteObject(*o*, *t*, *d*), it may have become a casualty link or a crucial surviving sibling link. A *casualty link* is one that has already been implicitly destroyed as the result of the attempted deletion of a related object in a link cycle—e.g., the `y2↔z1` link of A2 for object `z1` in Fig. 10.10, Case 1, when A1 is processed first. A *vital surviving sibling link* is one that has had a sibling link, i.e., one having the same related (parent) object, destroyed and is now vital in satisfying a lower bound on the number of siblings required—e.g., the `y2↔z1` link of A2 for object `y2` in Fig. 10.13, when the `x1↔y1` of A1 is processed first and the sibling link `y1↔z1` as been destroyed. The later link will be called a *sacrificial sibling link.*

Without link cycles, an entry link does not change, and there are no casualty links or vital surviving sibling links. With link cycles, whether or not a specific link is an entry or casualty link or has become a vital surviving sibling link before its processing in DeleteObject(*o*, *t*, *d*) is processing order dependent. Therefore, to show processing order independence, we must show that *R* will be unaffected if any link, $o \leftrightarrow o_k$, $1 \leq k \leq n$, has become an entry or casualty link, i.e., it has already been destroyed, or has become a vital surviving sibling link, rather than a sacrificial one, before being processed by DeleteObject(*o*, *t*, *d*). We consider below for each component of *R* each possible implicit destructibility binding for the object class of *o* in the association of which $o \leftrightarrow o_k$ is a link. This binding is given as *impB* in DeleteObject (Fig. 10.6) and as *b* in EnforceBinding (Fig. 10.4), which is invoked by DeleteObject and shows the actions taken in each case of implicit binding.

Exit with exception. The only situation in which DeleteObject exits with an exception is when a |- is detected by EnforceBinding. This occurs in case "|-" when detected in the immediate invocation of DeleteObject and cases "|?" and "|!" when detected in a recursive invocation.

First, assume the |- binding. Here we apply the theorem's hypothesis that there are no cyclic one-ended |- bindings, which means that any |- binding for links within a link cycle is two-ended. If $o \leftrightarrow o_k$ has already been destroyed, then an exception will have already occurred because of the |- binding for the o_k object class. (This situation is exemplified in Fig. 10.10, Case 1 where the implicit default binding for class Y in association A2 is replaced with a |-.) Thus, with a |- binding at both ends of the association, the link $o \leftrightarrow o_k$ cannot really be an entry or casualty link. It also cannot be a vital surviving sibling link or a sacrificial sibling link since no sibling link can be implicitly destroyed.

Now assume the |? binding and an *sLB* violation or a |! binding, and assume that the invocation of DeleteObject on o_k results in an exception. If $o \leftrightarrow o_k$ has already been destroyed, then a DeleteObject has already been invoked on o_k, which has resulted or will result in the same exception (based on encountering a |- binding unrelated to $o \leftrightarrow o_k$). Note that if o_k has been implicitly deleted as a result of a ' binding, is uncommitted, and will subsequently be undone (i.e., we are in an *nT* transaction that will be aborted), then the results of DeleteObject(*o*, *t*, *d*) will also be undone.

t.Destroys. In all cases of *impB*, $o \leftrightarrow o_k$ is destroyed in DeleteObject before EnforceBinding is called. Thus, *t.Destroys* is unaffected if $o \leftrightarrow o_k$ has already been destroyed or is or is not a vital surviving sibling link.

t.Deletes'. In all cases of *impB*, *o* is added to *t.Deletes*, and thus to *t.Deletes'*, in DeleteObject before EnforceBinding is called. Cases "|?", "|!", and "|'" of *impB* may implicitly delete o_k, thus adding it to *t.Deletes'*; however, if $o \leftrightarrow o_k$ has already been destroyed, then a DeleteObject has already been invoked on object o_k, and o_k is already in *t.Deletes'*.

If $o \leftrightarrow o_k$ has not already been destroyed, it may be a vital surviving sibling link, which means that a sibling link has already been destroyed, i.e., sacrificed. If so, then o_k has not been deleted, i.e., it is not in *t.Deletes'*. Cases "|!", and "|'" of *impB* always implicitly delete o_k, thus adding it to *t.Deletes'*. This is true even if $o \leftrightarrow o_k$ is not a vital surviving sibling link. But if it is, for case "|?", *sLB* is violated and o_k is implicitly deleted, thus adding it to *t.Deletes'*. This result is different than that which would have occurred had $o \leftrightarrow o_k$ not been a vital surviving sibling link, but instead a sacrificial sibling link, i.e., one whose destruction did not violate *sLB*. In this case, o_k is not implicitly deleted and thus not added to *t.Deletes'*. Regardless, whether $o \leftrightarrow o_k$ is the vital surviving sibling link or it is merely a sacrificial one and some other link is the vital surviving sibling link, o_k, the common related object, will be implicitly deleted and added to *t.Deletes'*. This situation is exemplified in Fig. 10.16 (a revision of Fig. 10.13) where on an attempt to explicitly delete `x1`, regardless of which link, `y1↔z1` or `y2↔z1`, becomes the sacrificial one and which the vital surviving one, `z1` is implicitly deleted.

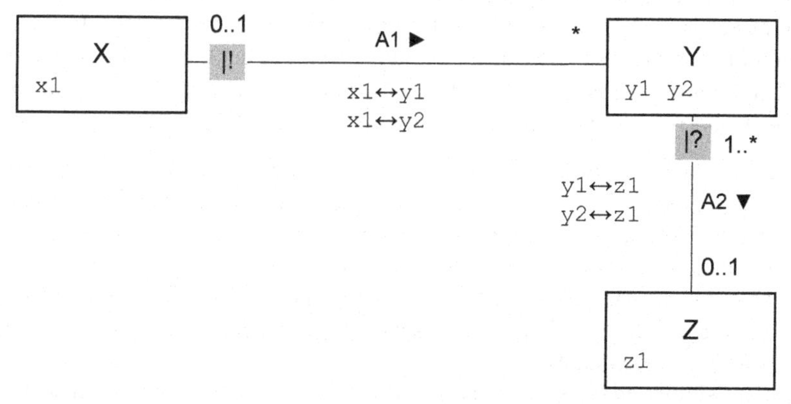

Fig. 10.16 Fig. 10.13 revised with a |! binding for class X in A1 and a |? binding for class Y in A2

***t.LbChecks'*.** Only cases "nil" and "|'" add (o_k, A^{-1}) to *t.LbChecks* if *sLB* is violated, where *A* is the association of which $o \leftrightarrow o_k$ is an instance.

If $o \leftrightarrow o_k$ has already been destroyed, then o_k is already in *t.Deletes'*, and thus it is immaterial that (o_k, A^{-1}) is not added to *t.LbChecks* since it won't be in *t.LbChecks'*, which is $\{(sO, A) \mid (sO, A) \in t.LbChecks \wedge sO \notin t.Deletes'\}$. We have already shown that the *t.Deletes'* component of *R* is unaffected by the order in which links are processed. (This situation is exemplified for *b* = "nil" in Fig. 10.10, Case 2.)

If $o \leftrightarrow o_k$ has not already been destroyed, it may be a vital surviving sibling link. If so, then o_k has not been deleted, i.e., it is not in *t.Deletes'*. If *sLB* is violated, then it is violated because of the loss of a sibling link and (o_k, A^{-1}) is added to *t.LbChecks*. For case "|'" (i.e., *impB* and *b* is "|'"), this is immaterial since object o_k is implicitly deleted and so (o_k, A^{-1}) will not be in *t.LbChecks'*. For case "nil", however, (o_k, A^{-1}) is essentially added to *t.LbChecks'*. This result is different than that which would have occurred had $o \leftrightarrow o_k$ not been a vital surviving sibling link, but instead a sacrificial sibling link, i.e., one whose destruction did not violate *sLB*. (This situation is exemplified in Figs. 10.13, 10.14, and 10.15 where either of the links for A2, shown in Fig. 10.13, can become the vital sacrificial sibling link depending on the order in which the links of A1 are processed on deletion of `x1`.)

Here, however, we apply the theorem's hypothesis that no cyclic non-dependent |' binding with possible partly sacrificial siblings is given for an association having links within a link cycle. This ensures that a |' binding was not the cause of the attempt to implicitly delete *o*, i.e., the invocation of DeleteObject(*o, t, d*), or if a |' binding was the cause, its associated multiplicity implied a dependency. This in turn ensures that regardless of whether $o \leftrightarrow o_k$ is a vital surviving link or a sacrificial link, the results will be the same: a lower bound violation will be detected by the Commit of the transaction that wraps the DeleteObject(*x, t, d*) as seen in Fig. 10.3 causing the complex object deletion to fail. Either the violation will be caused by the deletion of a vital surviving $o \leftrightarrow o_k$ link or, in the case of a |' binding with a dependent multiplicity (imagine the 0..1 multiplicity for class X in Fig. 10.13 was 1), it will be caused by the violation of this dependency. (That is, the situation won't be like that shown in Section 10.2.2.2 where a nested transaction commit detects a lower bound violation in trying to implicitly delete *o*, here `y1` or `y2`, but the exception is ignored. The nested transaction is simply aborted, as seen by the "Commit(*nT, d*); If exception the Abort *nT*;" in the pseudocode for case "|'" in EnforceBinding, Fig. 10.4.)

Since we have shown that for all components of *R*, the result of executing DeleteObject(*o, t, d*), where *o* is *x* or any related object of *x*, are unaffected by the order in which links are processed, the theorem is proved for explicit object deletion.

Now, we consider explicit link destruction and link change.

If a link is being explicitly destroyed or changed and if no object in the link is implicitly deleted, only one link is processed. Thus no dependency on the order in which links are processed exists.

To more simply address the situation where an object in an explicit link destruction or change is implicitly deleted, we examine equivalent operations involving explicit object deletion. An explicit destruction of or change in a link $x \leftrightarrow y$ in an association *A* will be viewed in terms of the explicit deletion of an object *a*, which represents the association instance, in a class *A*, which represents the association. The association *A* with link $x \leftrightarrow y$ is modeled in Fig. 10.17 (a) and remodeled in Fig. 10.17 (b) as a class *A* with object *a*. The multiplicities for classes *X* and *Y*, m_X and m_Y, have been mapped to equivalent multiplicities on class *A*, and the explicit cascade bindings for classes *X* and *Y*, b_X and b_Y have been mapped to equivalent implicit bindings on class *A*.

Based on the algorithms given in Section 10.1, we first show that explicitly destroying the link $x \leftrightarrow y$ in the model in Fig. 10.17 (a) is equivalent in outcome to explicitly deleting the object *a* in the model in Fig. 10.17 (b). The outcome of DestroyLink($x \leftrightarrow y$, *t, d*), as can be seen in Fig. 10.7, is determined by invoking EnforceBinding(b_X, *A, x, y, t, d*), where the multiplicity for the subject class in *A* is m_X, and EnforceBinding(b_Y, A^{-1}, *y, x, t, d*), where the multiplicity for the subject class in A^{-1} is m_Y. The outcome of DeleteObject(a, t, d), as can be seen in Fig. 10.6, is determined by invoking EnforceBinding(b_X, A_Y, a, y, t, d), where the multiplicity for the subject class in A_Y is m_X, and EnforceBinding(b_Y, A_X, a, x, t, d), where the multiplicity for the subject class in A_X is m_Y. Since the outcome of the EnforceBinding algorithm is determined only by the subject class binding b, the subject class multiplicity (which determines "if sLb is violated"), and the related object rO, the results of these EnforceBinding executions in both DestroyLink and DeleteObject algorithms are the same.

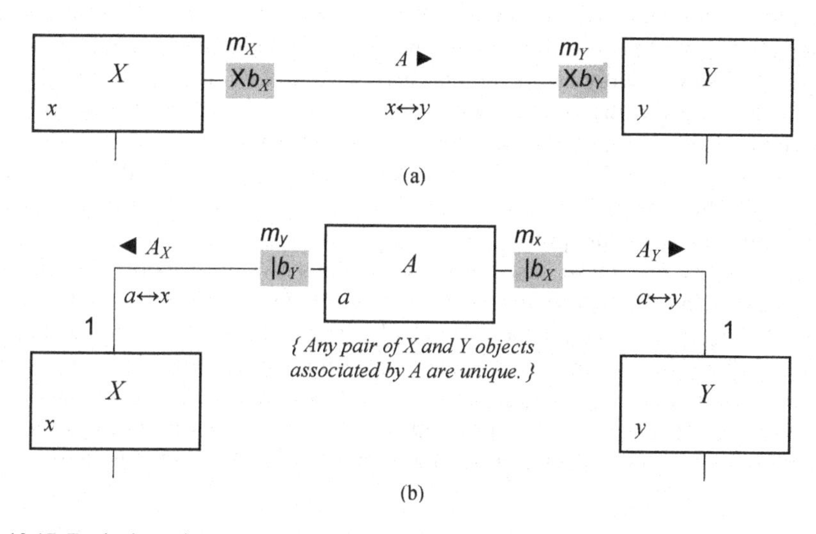

Fig. 10.17 Equivalence between (a) explicit destruction of an A link and (b) deletion of an A object

There are, however, two differences in these algorithms. In DestroyLink the order in which the two EnforceBinding invocations occur is always the same and the second invocation is guaranteed. While in DeleteObject, the order of these invocation is indeterminate and the second execution may not take place if a link cycle causes one of the links to be a casualty link. We have already shown that the order in which links are processed in DeleteObject is immaterial, thus we can assume in our analysis here that they are processed in the same order as that done in DestroyLink. Now, suppose that the second EnforceBinding is invoked in DestroyLink, i.e., EnforceBinding(b_Y, A^{-1}, y, x, t, d), when its equivalent in DeleteObject would not be invoked. This means that object x has already been implicitly deleted. In this case, the invocation of EnforceBinding is a no-op as a study of the case statement reveals. The "insert(x, A) into t.LbChecks" is irrelevant for nil and "|" since x is in $t.Deletes'$. A "-" is impossible since the theorem hypothesis does not allow a cyclic one-ended - binding (and a two-ended one would have already resulted in an exception). Finally, any DeleteObject(x, t, d) is a no-op since x is in $t.Deletes'$.

Now, since we have shown the equivalence of the destruction of $x \leftrightarrow y$ and the explicit deletion of a and since we have already proved the theorem for explicit deletion, the theorem is also proved for explicit link destruction.

Finally, we complete the proof of the theorem by addressing link change when the object to be replaced in the link must be implicitly deleted. We do this by showing that the result of executing the ChangeLink algorithm invoked in a nested transaction t to change a link $x \leftrightarrow y_1$ to $x \leftrightarrow y_2$ in database d is unaffected by the order in which the links are processed.

As can be seen in Fig. 10.8, the result of executing ChangeLink($x \leftrightarrow y_1$, y_2, t, d) is determined first by the task of implicitly destroying the $x \leftrightarrow y_1$ link and enforcing the explicit binding for the class of x in the applicable association and second by the task of attempting to implicitly create the link between x and y_2. If the result of the first task is independent of the order in which links are processed, then so is the result of the second. This is because the operations to ensure that objects x and y_2 still exist and, if so, to create the $x \leftrightarrow y_2$ link are clearly unaffected by the order of any link processing. Thus, we need only show that the result of the first task is unaffected by the order in which links are processed.

To do this, we first show, based on the algorithms given in Section 10.1, that implicitly destroying the link $x \leftrightarrow y_1$ and enforcing the Xb_X bind in the model in Fig. 10.18 (a) is equivalent in outcome to explicitly deleting the object a_1 in the model in Fig. 10.18 (b), assuming the link $a_1 \leftrightarrow x$ has been implicitly destroyed without enforcing the |b_Y binding prior to enforcing the |b_X binding. This assumption is valid since the "Destroy link $x \leftrightarrow y_1$;" in ChangeLink($x \leftrightarrow y_1$, y_2, t, d) totally destroys the connection between x and y_1 and ChangeLink does not enforce the Xb_Y binding.

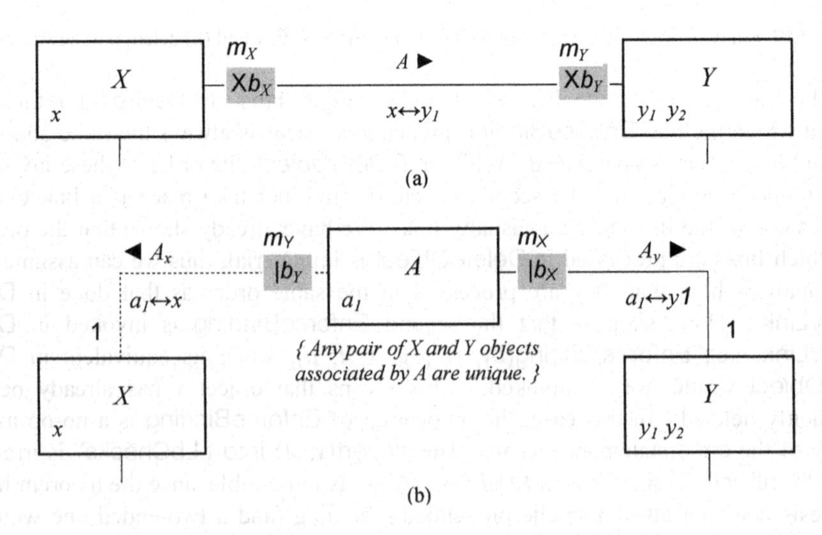

Fig. 10.18 Equivalence between (a) explicit change of an A link and (b) deletion of an A object

The outcome of implicitly destroying the link $x \leftrightarrow y_1$ in ChangeLink($x \leftrightarrow y_1$, y_2, t, d), as can be seen in Fig. 10.8, is determined by invoking EnforceBinding(b_X, A, x,

y_1, t, d) where the multiplicity for the subject class in A is m_X. The outcome of DeleteObject($a1$, t, d), as can be seen in Fig. 10.6, is determined solely by invoking EnforceBinding(b_X, A_Y, a, y_1, t, d), where the multiplicity for the subject class in A_Y is m_X, again assuming that $a_1 \leftrightarrow x$ is implicitly destroyed without enforcing the b_Y binding. This outcome is equivalent to that which would occur if link $a_1 \leftrightarrow x$ did not exist. Since the outcome of the EnforceBinding algorithm is determined only by the subject class binding b, the subject class multiplicity (which determines "if *sLb* is violated"), and the related object *rO*, the results of the EnforceBinding execution in both the ChangeLink and DeleteObject algorithms are the same.

Therefore, since we have shown that the implicit link destruction of $x \leftrightarrow y_1$ is equivalent to the explicit deletion of a_1 assuming no $a_1 \leftrightarrow x$ link, since we have already proved the theorem for explicit deletion, and since the "implicit link creation" part of link change is independent of link processing order, the theorem is now proved for link change.

Since the theorem has been proven for link creation, explicit object deletion, explicit link destruction, and link change, the theorem is now proven for all complex object operations. ■

10.3 Conclusion

This chapter has presented algorithms that can be used to implement ORN. They have been described at a level of abstraction that is independent of the type of database system, object or relational. The algorithms have been successfully implemented in OR+, an object database system, and ORN Additive, a relational database system.

In this chapter, we have also explored the problems posed by link cycles. We have seen how circularity is avoided by the detection of such cycles in the given algorithms and that the results of these algorithms are largely predictable in their presence. That is, complex object operations are non-circular and independent of the order in which association links are processed, except for two problematic ORN specifications.

One of these is the cyclic one-ended |- binding, which is given for an association having links within a link cycle. This specification **may** cause processing order dependencies. The one-ended |- binding is equivalent to the RESTRICT referential action in relational databases, which is similarly problematic.

The other problematic specification is the cyclic non-dependent |' binding with partly sacrificial siblings, which again **may** cause processing order dependencies in link cycles. The |' binding has no equivalent referential action in relational databases; however, if similar semantics were implemented by triggers, similar ambiguities could result.

Possible solutions to avoid the ambiguities caused by the two problematic specifications were discussed and, more importantly, a theorem was stated and proved that in the absence of these two specifications, ORN semantics are unambiguous.

Chapter 11

Adding ORN to the SQL Standard for RDBMSs

The current standard for RDBMSs is SQL:2008 (ANSI 2008). This chapter, based on Ehlmann (2006), describes how ORN can be incorporated into standard SQL.[1]

The chapter is organized into four sections. Section 11.1 gives the motivation for adding ORN to SQL. It reiterates some of the discussion given in Chapter 1 on the problems of modeling associations and implementing them in relational databases. Section 11.2 provides an overview of the existing capabilities for defining associations in SQL:2008. It reviews some material provided in Chapter 1 but also includes object-oriented, association capabilities that have been added to standard SQL. Section 11.3 gives the proposed syntax, semantics, and pragmatics for extending SQL with ORN. Section 11.4 concludes by highlighting the benefits of this extension.

11.1 Motivation

Despite the importance of relationships in database development, their translation from modeled representations into implementation has not been straightforward. That is, the way relationships are modeled during requirements analysis—often via an ER or class diagram—does not match the way they are defined to a DBMS—often via SQL. For example, a relationship modeled as one-to-many in an ER diagram must be defined as a foreign key minus a UNIQUE constraint in SQL; the one-to-many designation is lost. In addition, more detailed knowledge about relationship semantics captured during analysis and documented clearly in extended ER or class diagrams often gets obscured during implementation and is difficult to implement. For example, the fact that an employee must be in a department, a precise multiplicity of 1 for the Department class in a class diagram, is disguised in SQL as a NOT NULL constraint on a foreign key. Similarly, the fact that a carpool's existence depends on at least two riders, a semantic easily modeled in a class diagram, is not so easily defined in SQL. It must be **implemented** as embedded code within application programs or as complex integrity constraints that trigger exceptions and deletion as appropriate.

In recent years much interest has developed in model-driven development (MDD), which according to Mellor et al. (2003) "offers the potential for automatic

---

[1] This work is based on an earlier work: "Incorporating Object Relationship Notation (ORN) into SQL—revised," in *Proc. ACM Southeast Regional Conf.*, 1-59593-315-8 © ACM, 2006. http://doi.acm.org/10.1145/1185448.1185535

B.K. Ehlmann, *Object Relationship Notation (ORN) for Database Applications*,
Advances in Database Systems 39, DOI 10.1007/978-0-387-09554-7_11,

transformation of high-level abstract ... models into running systems". MDD enables the reuse of expert knowledge, enhances quality as models are successively refined into running systems, reduces costs by automating this process, and increases the longevity of the software that results. These benefits can only be achieved when required mappings between models (e.g., association notations in class diagrams and SQL) and running systems (e.g., database systems) are straightforward and automated (Mellor et al. 2003).

11.2 Overview of SQL Association Capabilities

Foreign keys and reference types are the two means for implementing associations in SQL. An association between tables is implemented by defining a foreign key or reference type within a table, the referencing table, whose values reference rows in a referenced table. An association link exists when the foreign key value in a row of the referencing table matches a unique key value in a row of the referenced table or, alternatively, when a reference type in a row of the referencing table uniquely identifies a row of the referenced table. An association between tables in SQL is either one-to-one or, from the viewpoint of the referencing table, many-to-one.

SQL:2008 provides declarative constraints for defining unique keys, foreign keys, and related association semantics. To simplify the overview of these constraints, we assume that all keys are single column and that all foreign keys reference primary keys. SQL provides more general constructs for specifying the same constraints when keys are compound and foreign keys reference non-primary, unique keys.

A REFERENCES clause on a column declares it a foreign key and guarantees referential integrity, which ensures that all non-null values in the column match a value in the primary key column or *self-referencing* column of the referenced table. Fig. 11.1 (a) gives the SQL needed to implement a many-to-one, or more precisely a *-to-0..1, association between employees and units using the foreign key unit defined in table Employee. unit references the primary key column id in table Unit.

```
CREATE TABLE Unit (
 id   CHAR(5)  PRIMARY KEY,
 ...
);
CREATE TABLE Employee (
 ssn  CHAR(11)  PRIMARY KEY,
 ...
 unit  CHAR(5)  REFERENCES Unit
);
```

(a)

```
CREATE TYPE UnitType AS (
 id   CHAR(5),
 ...
);
CREATE TABLE Unit OF UnitType (
 REF IS unitRef SYSTEM GENERATED
);
CREATE TABLE Employee (
 ssn   CHAR(11)  PRIMARY KEY,
 ...
 unit  REF(UnitType) SCOPE(Unit)
       REFERENCES Unit(unitRef)
);
```

(b)

Fig. 11.1 Association implemented using (a) a foreign key and (b) a reference

Fig. 11.1 (b) gives the SQL needed to implement the same association using the reference type unit defined in table Employee. The column unit references the self-referencing column unitRef in the typed table Unit. Here, the association could be implemented without the REFERENCES clause, i.e., the reference type need not be declared as a foreign key; however, referential integrity would then not be guaranteed, which could result in dangling references (Melton 2003, Melton J personal communication).

Placing NOT NULL and UNIQUE constraints on a foreign key or reference type implements certain association multiplicities. For example, in Fig. 11.1 (a) placing a NOT NULL constraint on unit, as shown below, constrains each employee to belong to a unit, implementing the *-to-1 association between employees and units.

```
unit CHAR(5) NOT NULL REFERENCES Unit
```

Placing a UNIQUE constraint on unit, as shown below, constrains each unit to be associated with at most one employee, implementing a 0..1-to-0..1 association between employees and units.

```
unit CHAR(5) UNIQUE REFERENCES Unit
```

Placing both a NOT NULL and UNIQUE constraint on unit would implement a 0..1-to-1 association between employees and units.

```
<references specification> ::=
   REFERENCES <table name> [ ( <self-referencing column name> ) ]
      [ ON DELETE <referential action> ] [ ON UPDATE <referential action> ]

<referential action> ::=
   NO ACTION | RESTRICT | SET NULL | SET DEFAULT | CASCADE
```

Fig. 11.2 Syntax of the existing REFERENCES clause

Some association semantics are defined by how the DBMS should react when deletions would cause referential integrity to be violated. Some desired actions can be specified as part of the REFERENCES clause. The syntax of this clause is given in Fig. 11.2. The ON DELETE phrase specifies the action to be performed when a row from a referenced table is deleted having matching rows (via foreign key values) in the referencing table. NO ACTION, the default, specifies that no referential action be taken, although if any unmatched foreign key values exist after the deletion (at transaction commit), the deletion is undone and an integrity constraint violation exception is raised. RESTRICT specifies that the deletion raise an exception immediately. SET NULL specifies that the matching foreign key values be set to NULL, and SET DEFAULT specifies that they be set to the specified default value for the foreign key. CASCADE specifies that all matching rows in the referencing table be deleted. The following definition of unit, if given in the Employee table, would cause all employees in a unit to be deleted when the unit is deleted.

```
unit CHAR(5) REFERENCES Unit ON DELETE CASCADE
```

Similar referential actions can be given for ON UPDATE; however, this clause has little to do with defining association semantics and more to do with key value maintenance and thus will not be discussed further.

When the declarative constraints discussed above are insufficient to implement association semantics, the general constraint capability of SQL can often be used—for example, to implement more precise multiplicities. The constraint below, if given in the Unit table in Fig. 11.1 (a), implements a 1..100-to-0..1 association between employees and units.

```
CONSTRAINT C1 CHECK
    ((SELECT COUNT(*) FROM Employee E
     WHERE E.unit = Unit.id) BETWEEN 1 AND 100)
```

While general constraints offer power and flexibility, they can become very complex and provide only the action of raising an exception when a constraint is violated.

```
CREATE TABLE Employee (
  ssn        CHAR(11) PRIMARY KEY,
  ...
  carpool  CHAR(8) REFERENCES Carpool ON DELETE SET NULL ON UPDATE CASCADE
);
CREATE TABLE Carpool (
  id          CHAR(8) PRIMARY KEY,
  ...
  CONSTRAINT C1 CHECK
  ((SELECT COUNT(*) FROM Employee WHERE carpool = Carpool.id) >= 2)
);
CREATE TRIGGER T1 AFTER DELETE ON Employee
  REFERENCING OLD ROW AS oldE FOR EACH ROW
  BEGIN ATOMIC
   IF (NOT(oldE.carpool IS NULL)) THEN
     IF ((SELECT COUNT(*) FROM Employee
          WHERE carpool = oldE.carpool) < 2) THEN
        DELETE FROM Carpool WHERE Carpool.id = oldE.carpool;
     END IF
   END IF
  END;
CREATE TRIGGER T2 AFTER UPDATE OF carpool ON Employee
  REFERENCES OLD ROW AS oldE
  REFERENCES NEW ROW AS newE
  FOR EACH ROW
  WHEN NOT(oldE.carpool IS NULL) AND -- †
            AND NOT(newE.carpool IS NULL AND
                 NOT EXISTS(SELECT * FROM Carpool
                              WHERE Carpool.id = oldE.carpool))
  BEGIN ATOMIC
   IF ((SELECT COUNT(*) FROM Employee
        WHERE carpool = oldE.carpool) < 2) THEN
      DELETE FROM Carpool WHERE Carpool.id = oldE.carpool;
   END IF
  END;
```

† - next three lines are needed to ensure foreign key is not being set to NULL as a result of a carpool deletion.

Fig. 11.3 SQL implementing employee-carpool association

This latter problem may be rectified by using triggers, which provide an even more powerful means to implement association semantics. Fig. 11.3 shows CREATE TRIGGER statements defined in the SQL that is required to implement the employee–carpool association as modeled in Figs. 1.24 (a) and 2.14. When triggers must be used, however, the SQL can become even more complex and error-prone. The processing order of referential actions, trigger executions, and constraint checking is governed by SQL and can result in subtle errors if care is not taken (e.g., see footnote in Fig. 11.3). Also, this ordering may not result in the desired association semantics. For example, using referential actions, triggers, and constraints to implement the '<0..1-to-*> association between units in conjunction with the <1-to-*> association between units and employees, as modeled in Figs. 1.24 (a) and 2.14 and discussed in Chapters 1 and 2, becomes very problematic in SQL (Ehlmann and Yu 2005a). We can remedy these problems by incorporating ORN into SQL.

11.3 Proposed ORN Extension to SQL

Extending SQL with ORN is relatively simple. Essentially, an *<association>*, as defined in Chapter 2, can be specified in place of the current REFERENCES clause for reference types and in place of the current ON DELETE phrase of the REFERENCES clause for foreign keys. The syntax, which redefines the *<references specification>* in SQL (from SQL:2003 specification), is given in Fig. 11.4.

```
<references specification> ::= <association> | <references clause>

<references clause> ::=
  REFERENCES < table name>
    [ <association definition> | ON DELETE <referential action> ]
    [ ON UPDATE <referential action> ]

<association definition> ::=
  <association> [ ON DELETE { SET NULL | SET DEFAULT } ]

<referential action> ::= NO ACTION | RESTRICT | SET NULL | SET DEFAULT | CASCADE
```

Fig. 11.4. Proposed syntax for *<references specification>*

The semantics for an *<association>* in a *<references specification>* are those given in Chapter 2, except that in the context of SQL, "class" translates to "table," "object" to "row," and "link" to a row reference, either via a foreign key or reference type. Table 11.1 describes ORN semantics in terms of the relational model. (The table is very similar to and consistent with Table 6.3.) In the relational model, a link is created when its implementing row reference is set to a reference a row, a link is destroyed when a non-NULL row reference is set to NULL, and a link is explicitly changed when a non-NULL row reference is set to reference another row.

In SQL, an *<association>* cannot be many-to-many, e.g., <*-to-1..*>. Such associations must, of course, be remodeled as two one-to-many associations before implementation in SQL (see Chapter 8, Section 8.1.1).

Table 11.1 ORN Semantics for the relational model

Semantics are given in terms of a subject table *S* with multiplicity *m* and binding *b* in an association *A* with a related table *R* (which could be table *S* in a different role).

S | b | m | A | ... | R

An *A* link is implemented by a foreign key reference, *fk*, from a row of *S* to a row of *R* or vice versa. An *A* link, now an "*A* reference," is destroyed when an *fk* reference is set to NULL or changed to reference a new row.

```
CREATE TABLE S (
  pk ... PRIMARY KEY,
  fk ... REFERNCES R (pk) b<m-to-...>...,
  ...
);
CREATE TABLE R (
  pk ... PRIMARY KEY,
  ...
);
```

or

```
CREATE TABLE S (
  pk ... PRIMARY KEY,
  ...
);
CREATE TABLE R (
  pk ... PRIMARY KEY,
  fk ... REFERNCES S (pk) <...-to- m>b ...,
  ...
);
```

<multiplicity>: Semantics are similar to those in UML. Essentially, *m* indicates a lower bound and upper bound on the number of rows of table *S* that reference via *fk* a particular row of table *R* (or the number of rows of table *S* that can be referenced via fk by a row of table *R*).

An *S* row can be created provided this does not violate *m*. The check for a lower bound violation is deferred until transaction commit. A *fk* can be set provided this does not violate *m*. The check for an upper bound violation is immediate. The enforcement of *m* on the deletion of an *S* row or on setting *fk* is determined by the binding *b*.

<binding>: A **|** in *b* denotes a "cut" and an Implicit, i.e., system initiated, destruction of an existing *A* reference that must occur on deletion of an *S* row. An **X** in *b* denotes a "cross out" and an eXplicit, i.e., user initiated, destruction of an *A* reference.[1]

An *S* row deletion and an explicit destruction of an *A* reference are *complex object operations*. Deletion of an *S* row succeeds only if all existing references involving that row are implicitly destructible. Also, deletion of an *S* row or destruction of an *A* reference succeeds only if all required implicit row deletions succeed.

<di>: A destructibility indicator in *b* specifies the destructibility of an *A* (i.e., *fk*) reference. The meaning of each indicator is given below. This meaning can alternatively be described by the actions taken on an attempt to destroy an *A* reference. These actions are given in brackets. If a <*di*> is given after a **|**, it applies to the implicit reference destruction; if given after an **X**, it applies to explicit reference destruction; and if given alone, it applies to both. If a <*di*> is not given, i.e., is nil, for implicit reference destruction, explicit reference destruction, or both, default destructibility applies to whichever.

nil — *Default destructibility*. A reference can be destroyed provided this does not violate *m*.[2] [Destroy the reference. If *m* is violated[2], raise an exception[3].]

- *Negative destructibility*. A reference cannot be destroyed. [Raise an exception.[3]]

~? or ? — *Conditional cascade destructibility*. A reference can be destroyed, but if this violates *m* (**?**), the destruction must be cascaded (~) to the related *R* row, i.e., this row must be implicitly deleted. [Destroy the reference. Delete the related *R* row**?** If *m* is violated, yes; else no.]

~! or ! — *Emphatic cascade destructibility*. A reference can be destroyed, but the destruction must be cascaded (~) to the related *R* row. [Destroy the reference. Delete the related *R* row!]

~' or ' — *Tentative (or qualified) cascade destructibility*. A reference can be destroyed, but an attempt must be made to cascade (~) the destruction to the related *R* row; however, this implicit *R* row deletion must be undone if it fails, but is required if and only if its undoing would violate *m*.[2] (Think of the **'** as a "pruned back !" or as a "qualifying footnote reference" on the cascade.) [Destroy the reference. Delete the related *R* row.' (' – If an exception occurs, undo the delete and then, if *m* is violated[2], raise an exception[3].)]

1 - An *fk* reference change done as a single operation that replaces an *S* row with another is not treated as an explicit link destruction relative to table *S* (but is relative to *R*) and is allowed if allowed by other multiplicities and bindings.

2 - The check for a lower bound violation is deferred until the end of the current complex object operation.

3 - The current complex object operation is undone.

```
CREATE TYPE UnitType AS (
  id        CHAR(5),
  parent    REF(UnitType),
  ...
);
CREATE TABLE Unit OF UnitType (
  REF IS unitRef SYSTEM GENERATED,
  parent WITH OPTIONS SCOPE Unit <*-TO-0..1>'
  ...
);
CREATE TABLE Employee (
  ssn       CHAR(11) PRIMARY KEY,
  ...
  carpool   CHAR(8) REFERENCES Carpool ?<2..15-TO-0..1> ON UPDATE CASCADE,
  unit      REF(UnitType) SCOPE(Unit) <*-TO-1>,
  ...
);
CREATE TABLE Carpool (
  id        CHAR(8) PRIMARY KEY,
  ...
);
```

Fig. 11.5. Extended SQL implementing model in Fig. 1.24 (a)

Fig. 11.5 gives the extended SQL that implements the associations modeled in Figs. 1.24 (a) and 2.14. The SQL in bold implements only the "belongs to" association between employees and customers and should be compared with the SQL in Fig. 11.3.

The precise semantics of the ?<2..15-TO-0..1> employee–carpool association in the context of Fig. 11.5 is defined by Table 11.1. The |? binding, implicit in the ? binding, applies on deletion of an Employee row and means (paraphrasing from Table 11.1): a "belongs to" reference can be implicitly destroyed, i.e., a carpool reference can be set to NULL, but if this violates the multiplicity 2..15, the related Carpool row must be implicitly deleted. The X? binding, also implicit in the ? binding, applies on the explicit destruction of a "belongs to" reference and means (again, from Table 11.1): a "belongs to" reference can be explicitly destroyed, i.e., carpool can be explicitly set to NULL or changed from a non-NULL value to another, but if this violates the multiplicity 2..15, the related Carpool row must be implicitly deleted. The multiplicity 2..15 is violated, for example, when carpool is set to NULL in one of the just two Employee rows referencing a particular Carpool row.

The implicit default binding for the Carpool table in the ?<2..15-to-0..1> association applies on deletion of a Carpool row and means (from Table 11.1): a "belongs to" reference can be implicitly destroyed provided this does not violate the multiplicity 0..1. The explicit default binding for the Carpool class applies on the explicit destruction of "belongs to" reference and means (from Table 11.1): a "belongs to" reference can be explicitly destroyed, i.e., carpool can be explicitly set to NULL, provided this does not violate the multiplicity 0..1. A 0..1 multiplicity is never violated by destroying a reference to a Carpool row.

Below are more of the association semantics that are implemented in Fig. 11.5. They are described both conceptually and, within brackets, in terms of SQL.

- If an employee [Employee row] is deleted, the link to the employee's unit is implicitly destroyed [the Employee row's unit reference is implicitly set to NULL (and then eliminated by the row deletion)] (default binding and * multiplicity).
- If a unit [Unit row] is deleted, all descendant units [Unit rows that directly or indirectly reference the deleted row via parent] are implicitly deleted; however, a unit is not deleted if it has any employees [if it is referenced by unit in any Employee row] (|' binding and implicit default binding with a 1 multiplicity).
- If a link between units is destroyed [if a parent reference is destroyed, e.g., set to NULL], the child unit and all descendant units [Unit rows that directly or indirectly reference the child via parent] are implicitly deleted; however, again, a unit is not deleted if it has any employees (X' binding and implicit default binding with a 1 multiplicity).

When a *<references specification>* is given as an *<association>*, the applicable column must be a reference type. This new syntax, as shown in Fig. 11.5 for unit in table Employee and parent in table Unit, eliminates the redundancy inherent in the current syntax needed to specify referential integrity for reference types. When the current *<references clause>* is given for either of these reference types (instead of the *<association>*), the "REFerences" indication and the referenced table name is specified twice and the self-referencing column is specified even though it is implicit. The *<column name>* WITH OPTIONS ... syntax for a typed table in SQL allows a *<references specification>*, a scope, and other constraints to be given for reference types defined in the user-defined type related to the table, e.g., parent in UnitType in Fig. 11.5.

When a *<references specification>* is given as a *<references clause>* that uses an *<association>*, semantic equivalents to the ON DELETE options NO ACTION, RESTRICT, SET NULL, and CASCADE are implicit (see Section 5.1.3). For example, NO ACTION is implicit in the <*-to-1> association specified for unit in Fig. 11.5. If after deletion of a Unit row any Employee rows exist that don't reference a Unit row, the deletion is undone and a multiplicity violation exception is raised. RESTRICT would be implicit if a <*-to-1>|- or <*-to-0..1>|- association had been specified for unit instead of <*-to-1>. SET NULL is implicit in any ...-to-0..1> association, e.g., that for carpool in table Employee in Fig. 11.5. CASCADE is implicit in any ...-to-1>|! or...-to-0..1>|! association.

Two alternative, yet quite useful, flavors of "CASCADE" are implicit when a |? or |' binding is used instead of the |! binding. The |' is given for unit in table Unit. The effectiveness of this cascade stems from allowing the complex object operation attempting the implicit deletion to continue without exception in cases when the implicit deletion can't be done but isn't required to enforce multiplicities. The |? is given for carpool in table Employee, but is given for the referencing table instead of the referenced table. Its effectiveness stems from allowing the cascade to be conditional based on whether or not a lower bound multiplicity is violated.

In an *<association definition>*, the semantics of the ON DELETE phrase are similar to those of the existing ON DELETE *<referential action>* phrase. Here, the phrase is only applicable to a ...-to-0..1> type of *<association>* given for a foreign key that is not a reference type. In the REFERENCES clause for carpool in Fig.

11.5, an ON DELETE SET NULL is implied by default and means that on deletion of a row in the Carpool table, all carpool values in the Employee table that reference the deleted row are set to NULL. If, however, ON DELETE SET DEFAULT was specified just after ?<2..15-to-0..1> in the REFERENCES clause, then these carpool values would be set to set to the specified default value for the column.

The remainder of this section discusses some of the finer points and impacts of incorporating ORN into SQL as described above.

Syntax to specify the *<referenced table and columns>* can replace *<table name>* in the *<references clause>* to allow the specification of a non-primary, unique key or to explicitly declare the primary key or self-referencing column of the referenced table.

The ON DELETE alternative to an *<association definition>* is retained in the extended SQL for upward compatibility only. An ORN-extended SQL can be completely upward compatible with existing SQL. When an *<association>* is not given for a foreign key or reference, existing SQL semantics apply.

The option to give an ON UPDATE phrase with an *<association>* is allowed to manage changes to symbolic primary key values.

If an *<association>* is given for a foreign key or reference type, then NOT NULL and UNIQUE constraints are implicit as described in Table 11.2.

Table 11.2 The NOT NULL and UNIQUE constraints implied for a row reference

<association>	NOT NULL	UNIQUE
m-to-0..1 where m is not 0..1 or 1	no	no
m-to-1 where m is not 0..1 or 1	yes	no
m-to-0..1 where m is 0..1 or 1	no	yes
m-to-1 where m is 0..1 or 1	yes	yes

A NOT NULL or UNIQUE constraint given with an *<association>* must be consistent with the given multiplicities.

Finally, any SQL statement that requires enforcement of ORN semantics as the result of any declared *<association>* becomes a complex object (or row) operation as described in Table 11.1. The processing of the operation in the relational DBMS must implicitly invoke proper updating of relevant supertables and subtables. It must also be properly integrated with trigger executions, constraint checking, and the non-ORN specified referential actions.

11.4 Conclusion

Incorporating ORN into SQL has a number of advantages. In general, it facilitates a more model-driven development approach, allowing association semantics modeled in UML class diagrams to be more directly translated into SQL and automatically enforced by the DBMS. More specifically and regardless of whether or not ORN is added to the SQL standard, RDBMS vendors should consider including ORN as an extended feature to their SQL for the following reasons.

- The multiplicities and bindings for an association as modeled are mapped directly into an *<association>* specified for the foreign key or reference type that implements the association in SQL. This holds true not only for one-to-one and one-to many associations, but also for many-to-many associations, *n*-ary associations, and "is a" relationships when these are modeled (or remodeled) as described in Chapter 8, Section 8.1.1.
- More powerful association semantics are easily declared in SQL and enforced by the DBMS. Such semantics enforce multiplicities; delete a row in a table (e.g., Carpool) when a given number of rows in the referencing table (e.g., Employee) no longer reference it, either because a row has been deleted from the referencing table or a foreign key (e.g., carpool) has been set to NULL or changed; or implicitly delete a row in a table (e.g., Unit) only when it is no longer being used, meaning there are no longer any **required** references to the row (e.g., unit references in EMPLOYEE).
- The semantics for associations implemented using either foreign keys or reference types are implemented in a consistent manner and, for reference types, implemented with minimal syntax.
- Database implementations are made easier and less error-prone in that there is less need for database developers to code, test, and maintain complex constraint and trigger specifications. Again, to appreciate this, the amount and complexity of the code needed to implement the employees-carpool association in existing SQL (see Fig. 11.3) need only be compared to that needed in the ORN-extended SQL (see Fig. 11.5).
- Adding ORN to SQL is feasible and relatively simple as shown by the development of ORN Additive (see Chapter 6).

Not all of the technical issues regarding the incorporation of ORN into SQL have been addressed in this chapter. For example, the statements needed to add RXC mode to standard SQL, as it was added to T-SQL via ORN Additive, were not presented. The chapter, however, has shown the feasibility of its incorporation and discussed the benefits that would result.

Chapter 12

Adding ORN to the ODMG Standard for ODMSs

An *object data management system* (ODMS) allows objects created and manipulated in an object-oriented programming language to be made persistent and provides traditional database capabilities like concurrency control and recovery to manage access to these objects. An ODBMS, one type of ODMS, stores the objects directly in an object database. An *object-to-database mapping* (ODM), another type of ODMS, stores the objects in another database system representation, usually relational (Cattel et al., 2000).

The de facto standard for ODMSs is ODMG 3.0 (Cattel et al. 2000). This standard defines an Object Model to be supported by an ODMG-compliant ODMS. This chapter describes how ORN can be added to the Object Model. It is an extended version of Ehlmann (2008).[1]

The chapter is organized into four sections. Section 12.1 briefly describes the ODMG Object Model and discusses the motivation for adding ORN to the model. Section 12.2 shows how the ODMG Object Definition Language (ODL), which defines the Object Model, can be extended with ORN syntax and describes ORN semantics in terms of the model. Section 12.3 discusses and illustrates algorithms that can be used to implement ORN semantics in an ODMS that is based on the extended Object Model. Section 12.4 provides concluding remarks.

12.1 Motivation

The Object Model defines the object semantics that can be specified to an ODMS. These semantics deal with how objects are named and identified and the properties and behavior of objects. They also deal with how objects can relate to one another.

In addition to supporting generalization/specialization relationships, the Object Model supports one-to-one, one-to-many, and many-to-many binary relationships between object types. These are the non-inheritance, or structural, types of relationships, termed *associations* in UML. For example, a one-to-many association between carpools and employees can be defined in the Object Model. A carpool object is defined so that it can reference many employee objects, and an employee object is defined so that it can reference at most one carpool.

[1] Portions reprinted from "Adding more support for associations to the ODMG Object Model," *ICSOFT 2006*, Software and Data Technologies, Springer-Verlag, CCIS 10:257-269. © Springer-Verlag Berlin Heidelberg 2008. With kind permission of Springer Science and Business Media.

B.K. Ehlmann, *Object Relationship Notation (ORN) for Database Applications*,
Advances in Database Systems 39, DOI 10.1007/978-0-387-09554-7_12,

The Object Model prescribes that the ODMS automatically enforce referential integrity for all defined associations. This means that if an object is deleted, all references to that object that maintain associations must also be deleted. This ensures that there are no such references in the database that lead to nonexistent objects.

What has just been described is the extent of support for associations in the Object Model. What is lacking is some additional, easily implementable support for associations that could significantly improve the productivity of developing object database systems and the reliability of those systems.

For example, the Object Model, like the relational model, does not support the specification of precise multiplicities. For example, the multiplicity for the Employee class in the carpool–employee association may be given as 2..15 in a class diagram, meaning that a carpool must be related to a minimum of two and a maximum of fifteen employees. Such association semantics, documented during conceptual database design, are often lost during logical database design and definition unless supported by the logical data model, here the Object Model. If not supported, to survive, they must be resurrected by the programmer during implementation and for object databases translated into cardinality checking and exception handling code within relevant create and update methods.

The Object Model also does not support association semantics that are equivalent to those supported in relational DBMSs via the REFERENCES...ON DELETE clause of the CREATE TABLE statement in SQL (ANSI 2008). Such semantics would, for instance, allow one to declare an association between objects such that if an object is deleted, all related objects would be automatically deleted by the ODMS, i.e., an ON DELETE CASCADE. For example, if a unit within a company were deleted, all subordinate units would be implicitly deleted. Such an association semantic is required for an ODMS to provide support for composite objects.

This chapter shows how the Object Model can be extended with ORN and provides the algorithms that an ODMG-compliant ODMS can use to implement the association semantics specified by ORN. The ORN extension is very straightforward, and the algorithms required are relatively simple. The end result is an enhanced Object Model that supports more powerful association semantics—in fact, more powerful than those supported by relational systems without having to code complex constraints and triggers. Extending models with ORN and providing the required mappings between them—UML class diagram to Object Model to ODMS implementation—facilitates a model-driven development approach (Mellor et al. 2003).

The benefits of this approach are a major improvement in the productivity of developing object database systems and an increase in their reliability. Productivity is improved when translations from class diagram models into object models are more direct and when programmers do not have to develop code to implement association semantics. Currently, many developers working on many database applications must implement, test, and maintain custom code for each type of association, often "reinventing the wheel." Reliability is increased when the ODMS is responsible for enforcing association semantics. Currently, developers sometimes fail to enforce these semantics or inevitably introduce errors into database applications when they do.

12.2 Adding ORN to ODL

This section first describes how associations are defined in ODL and then shows how ORN syntax and semantics can be integrated into ODL.

12.2.1 Associations in ODL

In ODL an association is defined by declaring a relationship *traversal path* for each end of the association. A traversal path provides a means for an object of one class to reference and access the related objects of a *target class* (which is the same class in an intra-class association). Access to many target class objects requires the traversal path declaration to include an appropriate collection type, usually a set or list, that can contain references of target class type. Access to a maximum of one target class object requires the declaration to include a reference of target class type. A traversal path declaration must also include the name of its *inverse traversal path.*

For example, the one-to-many relationship between carpools and employees, discussed earlier and modeled by the class diagram in Fig. 12.1, would be declared in ODL as shown in Fig. 12.2. The 2..15 multiplicity given in the class diagram must be implemented by application code.

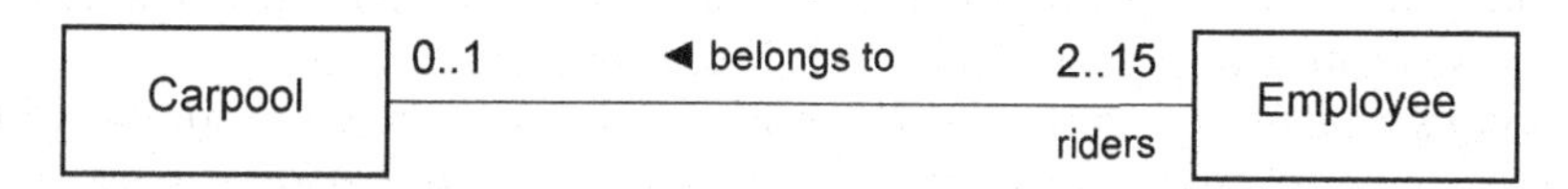

Fig. 12.1 Class diagram for carpool–employee association

```
class Carpool {
  ...
  relationship set<Employee>  riders
                              inverse Employee::carpool;
  ...
};
class Employee {
  ...
  relationship Carpool        carpool
                              inverse Carpool::riders;
  ...
};
```

Fig. 12.2 ODL for carpool–employee association

12.2.2 Adding ORN syntax

Adding ORN syntax to the Object Model is relatively straightforward. Essentially, ODL is extended to allow an <association> to be given for each declared relationship. The syntax for an <association> is the same as that given in Chapter 2. The ORN-extended ODL syntax is given in Fig. 12.3.

```
<rel_dcl>                  ::= relationship
                               <target_of_path> <identifier>
                               inverse <inverse_traversal_path>
                               [ <association> ]
<target_of_path>           ::= <identifier>
                               | <coll_spec> < <identifier> >
<inverse_traversal_path> ::= <identifier> :: <identifier>
```

Fig. 12.3 Updated BNF for a relationship in ODL

To illustrate the syntax and semantics of ORN in the context of the Object Model, a database containing the carpool–employee association as well as two other associations is modeled by the class diagram in Fig. 12.4, which is the same as Figs. 1.24 (a) and 2.14. The model is implemented by the ORN-extended ODL given in Fig. 12.5.

If an <association> is not given for a relationship in ODL (see Fig. 12.3), the default <association> is <0..1-to-0..1> for a one-to-one relationship, <0..1-to-*> for a one-to-many, and <*-to-*> for a many-to-many. These defaults give relationships the same semantics as they have in the existing Object Model.

An <association> given for a relationship need only to be given for one of the traversal paths. If given for both traversal paths, the <association>s must be inverses of each other. For example, an <association>, if given for carpool in Fig. 12.5, must be given as ?<2..15-to-0..1>.

When an <association> is given for a traversal path *tp* in class *C*, the multiplicity and binding given after the -to- apply to *tp* and to the target class, while the multiplicity and binding given before the -to- apply to the inverse *tp* and to class *C*. For example, in Fig. 12.5, the multiplicity 2..15 and binding ? apply to the traversal path riders and class Employee, and the multiplicity 0..1 and default bindings apply to the traversal path carpool and the target class Carpool. If the multiplicity given for a traversal path in an <association> implies "many," then the type of that traversal path must be a collection type.

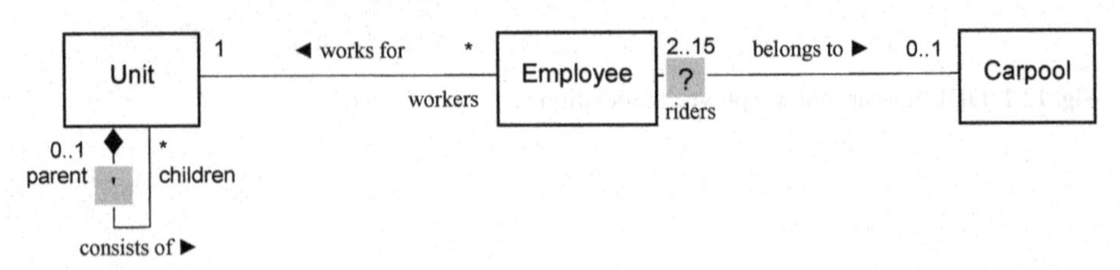

Fig. 12.4 ORN-extended UML class diagram

```
class Carpool {
  relationship set<Employee>  riders
                              inverse Employee::carpool <0..1 -to-2..15>?;
  ...
};
class Employee {
  relationship Carpool        carpool inverse Carpool::riders;
  relationship Unit           unit
                              inverse Unit::workers <*-to-1>;
  ...
};
class Unit {
  relationship set<Employee>  workers
                              inverse Employee::unit;
  relationship Unit           parent
                              inverse Unit::children;
  relationship set<Unit>      children
                              inverse Unit::parent '<0..1-to-*>;
  ...
};
```

Fig. 12.5 ODL for class diagram shown in Fig. 12.4

The last issue to address in extending ODL is association inheritance. In the Object Model, a relationship can be inherited by a class via the extends relationship. For example, the declaration class SalesPerson extends Employee { ... } would mean that the SalesPerson class inherits the attributes, relationships, and behavior of the Employee class. Thus, the carpool traversal path as declared in the Employee class in Fig. 12.5 would be inherited by the SalesPerson class, allowing sales people to belong to carpools. When a relationship is inherited by a class, all of the semantics defined by its <*association*>, given or defaulted, are also inherited.

And, of course, the semantics of all <*association*>s defined in the ODL—defaulted, given, or inherited—must be maintained by the ODMS.

12.2.3 Adapting ORN semantics to ODL

Table 2.1 described ORN semantics conceptually in terms of ER and class diagrams. Table 12.1 describes ORN semantics in terms of the Object Model, or ODL. (The table is very similar to and consistent with Table 7.1.) In the Object Model, instead of "association links" being "created" and "destroyed," "relationship references" (or "traversal path references") are "formed" and "dropped." Dropping a relationship or traversal path reference also means dropping the corresponding inverse reference (in the inverse traversal path). Also, bindings and multiplicities are now associated with traversal paths as well as with related classes. This is convenient for identifying bindings and multiplicities in intra-class associations since the subject class and related class, called the "target class" in ODL, are the same. Traversal path names can be equated to role names given in UML class diagrams. In Table 12.1, traversal path names tp_R and tp_S are also role names in the class diagram for relationship A.

Table 12.1 ORN Semantics for the Object Model

Semantics are given in terms of a subject class *S* with multiplicity *m* and binding *b* in a relationship *A* with a target, or related, class *R* (which could be *S* in a different role).

S | m b tp_S — A — tp_R | R

A is implemented by a traversal path $S::tp_R$ and inverse traversal path $R::tp_S$. The type of tp_R is *R* or some suitable collection of *R* (shown here as *R*), and the type of tp_S is *S* or some suitable collection of *S* (shown here as **set<S>**). The multiplicity for tp_S is *m*, and the binding is *b*. An *A* link, now an "*A* reference pair," is destroyed when a tp_S or tp_R reference is dropped or changed to reference a new object. Forming or dropping either traversal path reference is equivalent to and includes forming or dropping, respectively, its inverse traversal path reference.

```
class S {
   ...
   relationship R tpR inverse R::tpS
                        b<m-to-...;
}

class R {
   ...
   relationship set<S> tpS
                 inverse S::tpR;
}
```

***<multiplicity>*:** Semantics are the same as those in UML. Essentially, *m* indicates a lower bound and upper bound on the number of objects of type *S* that can be related via *A* to each object of type *R*.

An *R* object can be created provided this does not violate *m*. The check for a lower bound violation is deferred until transaction commit. A tp_R reference can be formed provided this does not violate *m*. The check for an upper bound violation is immediate. The enforcement of *m* on the deletion of an *S* object or dropping of a tp_R (or tp_S) reference is determined by the binding *b*.

***<binding>*:** A **|** in *b* denotes a "cut" and an Implicit, i.e., system initiated, idropping of an existing *A* reference that must occur on deletion of an *S* object. An **X** in *b* denotes a "cross out" and an eXplicit, user initiated, dropping of an *A* reference.[1]

An *S* object deletion and an explicit dropping of an *A* reference are *complex object operations*. Deletion of an *S* object succeeds only if all existing relationship references involving that object are implicitly destructible, i.e., can be dropped. Also, deletion of an *S* object or destruction of an *A* reference succeeds only if all required implicit object deletions succeed.

***<di>*:** A destructibility indicator in *b* specifies the destructibility of an *A* (i.e., tp_R and inverse tp_S) reference. The meaning of each indicator is given below. This meaning can alternatively be described by the actions taken on an attempt to destroy an *A* reference. These actions are given in brackets. If a *<di>* is given after a **|**, it applies to implicit reference destruction; if given after an **X**, it applies to explicit reference destruction; and if given alone, it applies to both. If a *<di>* is not given, i.e., is nil, for implicit reference destruction, explicit reference destruction, or both, default destructibility applies to whichever.

- nil — *Default destructibility*. A reference can be destroyed provided this does not violate *m*.[2] [Drop the reference. If *m* is violated [2], raise an exception [3].]
- **-** — *Negative destructibility*. A reference cannot be destroyed. [Raise an exception.[3]]
- **~? or ?** — *Conditional cascade destructibility*. A reference can be destroyed, but if this violates *m* (**?**), the destruction must be cascaded (**~**) to the related *R* object, i.e., this object must be implicitly deleted. [Drop the reference. Delete the related *R* object**?** If *m* **is violated**, yes; else no.]
- **~! or !** — *Emphatic cascade destructibility*. A reference can be destroyed, but the destruction must be cascaded (**~**) to the related *R* object. [Drop the reference. Delete the related *R* object**!**]
- **~' or '** — *Tentative (or qualified) cascade destructibility*. A reference can be destroyed, but an attempt must be made to cascade (**~**) the destruction to the related *R* object; however, this implicit *R* object deletion must be undone if it fails, but is required if and only if its undoing would violate *m*.[2] (Think of the **'** as a "pruned back !" or as a "qualifying footnote reference" on the cascade.) [Drop the reference. Delete the related *R* object.**'** (**'** – If an exception occurs, undo the delete and then, if *m* is violated [2], raise an exception [3].)]

1 - A tp_S reference change done as a single operation that replaces an *S* object with another is not treated as an explicit link destruction relative to class *S* (but is relative to *R*) and is allowed if allowed by other multiplicities and bindings.
2 - The check for a lower bound violation is deferred until the end of the current complex object operation.
3 - The current complex object operation is undone.

Table 12.1 provides the binding semantics of the ?<2..15-to-0..1> association between employees and carpools. The |? binding, implicit in the ?, applies on deletion of an Employee object and means (paraphrasing from Table 12.1): a "belongs to" reference can be implicitly destroyed, i.e., a carpool or riders reference (see Fig. 12.5) can be implicitly dropped, but if this violates the multiplicity 2..15, the related Carpool object must be implicitly deleted. The X? binding, also implicit in the ?, applies on the explicit dropping of a carpool or riders reference and means (again, from Table 12.1): a "belongs to" reference can be explicitly destroyed, i.e., a carpool or riders reference can be explicitly dropped, but if this violates the multiplicity 2..15, the related Carpool object must be implicitly deleted. The multiplicity 2..15 is violated when a reference to one of just two employees in a carpool, i.e., one of just two references in the set riders, is dropped.

The implicit default binding for the Carpool class in the ?<2..15-to-0..1> association applies on deletion of a Carpool object and means (from Table 12.1): a "belongs to" reference can be implicitly destroyed, i.e., a carpool or riders reference can be implicitly dropped, provided this does not violate the multiplicity 0..1. The explicit default binding for the Carpool class applies on the explicit dropping of a carpool or riders reference and means (from Table 12.1): a "belongs to" reference can be explicitly destroyed, i.e., a carpool or riders reference can be explicitly dropped, provided this does not violate the multiplicity 0..1. The 0..1 multiplicity is never violated by dropping a reference in riders or a carpool reference.

Below are more of the association semantics that are modeled in Fig. 12.4 and implemented in Fig. 12.5. They are described both conceptually and, within brackets, in terms of the Object Model.

- If an employee [Employee object] is deleted, the link to the employee's unit is implicitly destroyed [the object's unit reference to its target Unit object is implicitly dropped] (default binding and * multiplicity).
- If a unit [Unit object] is deleted, all descendant units [Unit objects recursively referenced via children] are implicitly deleted; however, a unit is not deleted if it has any employees [if workers references any Employee objects] (|' binding and implicit default binding and 1 multiplicity).
- If a link between units is destroyed [if a children reference (or its inverse parent reference) is dropped], the child unit and all of its descendant units [Unit objects recursively referenced via children] are implicitly deleted; however, again, a unit is not deleted if it has any employees (X' binding and implicit default binding and 1 multiplicity).

12.3 Algorithms

The implementation of ORN semantics in an ODMG-compliant ODMS is described by giving the algorithms required to create and delete objects and form, drop, and change relationship references. These operations become complex object operations in the context of ORN. This means they may no longer involve just one object or relationship reference but may involve many objects, relationships, and relationship references in the scope of a complex object.

In this section, the algorithms for these operations are given by providing related pseudocode, with commentary, for the ObjectFactory::new() and Object::delete() methods, which are associated with an object, and the *C*::form_*tp*(), *C*::drop_*tp*(), and *C*::change_*tp*() methods, which are associated with a declared traversal path *tp* in a user-declared class *C*. These methods, except for *C*::change_*tp*(), are defined as part of the Object Model (see Chapter 2 of Cattel et al. (2000)).

The algorithms have been developed by reverse engineering the code for implementing ORN within OR+ (see Chapter 7). This is the same code executed when one uses the ORN Simulator (see Chapter 3). Thus, the algorithms are well-tested.

Their implementation of ORN semantics is unambiguous in the presence of association cycles as long as <*association*>s do not contain a problematic, cyclic |- or |' binding, both of which are thoroughly explained in Section 10.2. Here, unambiguous means that the results of a complex object operation are independent of the order in which traversal paths and the references in these paths are processed. This property of ORN is discussed in detail and proven in Chapter 10.

As stated in Section 12.1, the algorithms are relatively simple; however, they depend on the ODMS implementation supporting a nested transaction capability. Nested transactions are needed to implement the semantics of the ' binding and are desirable so that the system can check multiplicity violations at the end of a complex object operation, undoing the operation upon any exception and thus making the complex object operation atomic. The Object Model defines a Transaction Model, which does not provide nested transactions. So, before giving the algorithms for the complex object operations, the Transaction Model is extended to support nested transactions, at least for the purpose of implementing ORN (but not necessarily as a user capability). Such support for nested transactions is assumed, and algorithms for transaction methods are given, focusing on the actions required to support ORN semantics.

All methods are assumed to execute in the context of an opened database *d*, and methods new(), delete(), form_tp_R(), drop_tp_R(), and change_tp_R() are assumed to execute within the scope of a user-defined transaction.

The pseudocode that expresses the algorithms is a mixture of ODL, C++, Java, and English and adheres as close as possible to the conventions of ODL. Indentation indicates compound statements within control structures, with appropriate end's often used to terminate these statements. The try...handle...end handle control structure for exception handling is similar to Java's try {...} catch {...}. Methods for a class are introduced with a header of the form Method <*variable*>.<*method name*>(...), where the <*variable*> is used in the body of the method to refer to the object on which the method is invoked, i.e., the implicit parameter and this object in C++ and Java. A <*method name*> begins with an underscore if it is to be invoked only by the ODMS implementation.

The algorithms are expressed using the variables defined in Table 12.1 and the following functions:

Type(*o*) – the type, or class, of object *o*, which is the most specific type of *o* in any type hierarchy.

LbM(*tp*) – the lower bound multiplicity for *tp* in the *<association>* for the relationship represented by traversal path *tp*.

UbM(*tp*) – the upper bound multiplicity for *tp* in the *<association>* for the relationship represented by traversal path *tp*.

ImpB(*tp*) – the implicit destructibility binding for *tp* (minus the | symbol) in the *<association>* for the relationship represented by traversal path *tp*.

ExpB(*tp*) – the explicit destructibility binding for *tp* (minus the X symbol) in the *<association>* for the relationship represented by traversal path *tp*.

Inverse(*tp*) – the inverse traversal path of *tp*.

Refs(*o.tp*) – the number of references in *o.tp*, which, if *tp* is a collection, is the cardinality of the collection, i.e., *o.tp*.cardinality() and, if *tp* is a reference, is 0 if nil and 1 if not.

12.3.1 Class **Transaction**

Fig. 12.6 gives a partial implementation of class Transaction, focusing on the code needed to implement ORN. Here, for simplicity the pseudocode departs some from the ODMG Transaction Model and adopts more the language of its ODMG C++ Binding. It is assumed that the static method current() returns the transaction currently in progress and that the proper constructor method is called to create a new transaction when a Transaction is declared. The second constructor, which requires a parameter, is invoked when declaring a nested transaction.

The variable *rxc_mode* records whether or not the transaction is in *Relationship eXChange(RXC) mode.* Setting this mode defers checks on lower bound multiplicities until the user-defined transaction commit so that normal ORN semantics, which could cause an exception or implicit delete of a target object, are not invoked. This permits "harmless" relationship exchanges within a transaction. For example, assume a database as defined in Fig. 12.5 where a Carpool object *c* references just two Employee objects, *e1* and *e2*. If a transaction dropped the reference from *c* to *e1*, *c* would normally be implicitly deleted (X? binding and multiplicity violation). But if RXC mode were set, the transaction could drop this reference and form another from *c* to an employee *e3*, for instance, without *c* being deleted. The RXC mode feature complements ORN. The method set_rxc_mode() sets the mode and reset_rxc_mode() resets it.

The set *deleted_objects* is used to record which objects will be deleted in this transaction or have already been deleted. The set *deferred_lbM_checks* is used to record which traversal paths in which objects must be checked at transaction commit for violations of lower bound multiplicity.

The begin() method initializes the RXC mode to false unless the transaction is nested, in which case it initializes the mode to that of the parent transaction. The method also initializes the sets *deleted_objects* and *deferred_lbM_checks* to empty. The statement "Perform other begin transaction functions" may involve invoking a "primitive" begin() transaction method. "Primitive" means the normal method that does not involve support for ORN. In the case of OR+, this is the begin() method on an associated Object Store transaction.

```
class Transaction {
private:
   static Transaction              current = nil;
   Transaction                     parent;  // nil iff not nested
   Boolean                         begun;
   Boolean                         rxc_mode;
   Set<Object>                     deleted_objects;
   Set<(Object, TraversalPath)>    deferred_lbM_checks;
   ...
public:
   Method t.Transaction()
   // Create a non-nested transaction t.
      parent = nil;  begun = false;

   Method t.Transaction(Transaction theParent)
   // Create a nested transaction t within Transaction
   //    theParent.
      parent = theParent;  begun = false;

   static Method Transaction current()
   // Returns the inner-most, current transaction.
      return current;

   Method t.begin() raises(TransactionInProgress,
                           DatabaseClosed)
   // Begin a transaction t.
      if begun then raise TransactionInProgress;
      if parent = nil then rxc_mode = false
      else rxc_mode = parent.rxc_mode;
      Set deleted_objects, deferred_lbM_checks to empty;
      Perform other begin transaction functions;
      begun = true;
      current = t;

   Method t.commit() raises(TransactionNotInProgress,
                            IntegrityError)
   // Commit transaction t.
      if not begun then raise TransactionNotInProgress;
      for each (object, traversal path (o, tp)) in
               deferred_lbM_checks do
         if o is not in deleted_objects or deleted_objects of
               any ancestor transaction then
            if Refs(o.tp) < LbM(tp) then raise IntegrityError;
      Perform other commit transaction functions;
      if parent ≠ nil then Include objects in deleted_objects
                              in parent.deleted_objects;
      begun = false;
      current = parent;

   Method t.abort() raises(TransactionNotInProgress)
   // Abort transaction t.
      if not begun then raise TransactionNotInProgress;
      Perform other abort transaction functions;
      begun = false;
      current = parent;

   Method t.set_rxc_mode()
            raises(TransactionNotInProgress)
   // Set Relationship eXChange (RXC) mode.
      if not begun then raise TransactionNotInProgress;
      rxc_mode = true;

   Method t.reset_rxc_mode()
            raises(TransactionNotInProgress)
   // Reset Relationship eXChange (RXC) mode.
      if not begun then raise TransactionNotInProgress;
      rxc_mode = false;

   Method t.in_rxc_mode()
            raises(TransactionNotInProgress)
   // Return true iff in eXChange (RXC) mode.
      if not begun then raise TransactionNotInProgress;
      return rxc_mode;

   Method t._check_path_at_commit(Object o,
                                   TraversalPath tp)
   // Ensure that traversal path o.tp is checked at commit
   //    of transaction for lower bound multiplicity violation.
      deferred_lbM_checks.insert_element((o, tp));

   Method t._check_path_at_parent_commit
               (Object o, TraversalPath tp)
   // Ensure that traversal path o.tp is checked at commit
   //    of the parent transaction for lower bound multiplicity
   //    violation.
      assert(parent ≠ nil);
      parent.deferred_lbM_checks.insert_element((o, tp));

   Method t._check_paths_at_commit(Object o)
   // Ensure that all traversal paths in object o with a lower
   //    bound multiplicity constraint are checked at commit
   //    of transaction for a lower bound multiplicity violation.
      deleted_objects.remove_element(o);
      for each traversal path tp in Type(o) do
         if LbM(o.tp) > 0 then
            deferred_lbM_checks.insert_element((o, tp));

   Method t._mark_for_deletion(Object o)
   // Mark object o for deletion in transaction t.
      deleted_objects.insert_element(o);

   Method boolean t._deleted(Object o)
   // Return true iff object o has been marked for deletion in
   //    this transaction or an ancestor transaction.
      if deleted_objects.contains_element(o) then
         return true
      else if parent = nil then return false
         else return parent._deleted(o);
   ...
};
```

Fig. 12.6 Definition of the Transaction class

The commit() method ensures that lower bound multiplicity constraints are satisfied for certain traversal paths in certain objects, i.e., certain *traversal path instances*. These include instances that have such constraints, i.e., a nonzero lower bound, and that 1) are in objects that have been created during the transaction (in this case, the user-defined transaction); 2) have had a reference explicitly dropped during the transaction (again, the user-defined transaction) while in RXC mode and the drop caused a lower bound multiplicity violation; and 3) have had references dropped implicitly or explicitly during the transaction (in this case, a system-defined, nested transaction) and require multiplicity checks at the end of the complex object operation because of their binding (see footnote 2 in Table 12.1). The required checks are done for each traversal path instance, i.e., object and traversal path pair, in the set *deferred_lbM_checks*, provided the object is not in the *deleted_objects* set of the current transaction or the *deleted_objects* set of any ancestor transaction. This provision ensures that checks are not made for objects that were or will be deleted in the transaction. An exception is raised if the number of references in a traversal path instance is less than the traversal path's lower bound multiplicity.

The statement "Perform other commit transaction functions" may involve invoking a "primitive" commit() transaction method. OR+ invokes the "primitive" commit() transaction method on the associated Object Store transaction.

The commit() method, if invoked on a nested transaction, includes its set of *deleted_objects* into the set of *deleted_objects* of its parent transaction before exiting.

In the abort() method, the statement "Perform other abort transaction functions" may involve invoking a "primitive" abort() transaction method. OR+ invokes the Object Store abort() method. This, of course, restores the database to the state it was in when the begin() was invoked on the transaction.

The remaining transaction methods defined in Fig. 12.6, are only invoked in the implementation of other methods and are discussed when they are first invoked.

12.3.2 Method new()

The new() method, given in Fig. 12.7, is described like the new operator in the ODMG C++ Binding. The method serves as a wrapper to invoke the _primitive_new() method, which refers to the currently defined new() method in the Object Model, which deals only with the object as a *primitive object*. The functionality that must be added to this method to support ORN must ensure that the object as a complex object is properly constructed. This means that all lower bound multiplicities for its traversal paths are satisfied before the transaction that invokes the new() can commit. To ensure this, new() invokes _check_ paths_at_commit(*o*) on the current transaction, where *o* is a reference to the newly created object.

```
Method Object fo.new(Database d, Class C)
// Create in database d and return a new object of type C from factory object fo.
  t = Transaction::current();
  o = fo._primitive_new(Database d, Class C);
  t._check_paths_at_commit(o);
  return o;
```

Fig. 12.7 Algorithm for the new() method in interface ObjectFactory

The _check_paths_at_commit() method, given in Fig. 12.6, first removes the object from the *deleted_objects* set (if a member). This step is included to address the situation where an ODMS assigns an object reference (or identifier) to a newly created object that was the same one used for an object that was previously deleted in the transaction. The method then inserts traversal path instances into the set *deferred_lbM_checks* for all traversal paths in the object that have lower bound multiplicity constraints. This must be done for traversal paths declared in the class Type(*o*) as well as those inherited.

12.3.3 Method delete()

The delete() method, seen in Fig. 12.8, is defined in the interface Object, which is implicitly inherited by all user-defined classes in the Object Model. The method replaces the delete() method as currently defined in Object Model. It provides the ORN semantics for the deletion of a complex object, yet retains the semantics of the currently defined delete() when all traversal paths of a class have default *<association>*s. The delete() method—like the form_tp_R(), drop_tp_R(), and change_tp_R() methods to be presented later—provides a nested transaction that embeds the complex object operation, permitting its effects on the database to be undone if an exception occurs. If one occurs, it is re-raised after aborting the transaction to allow the user-defined transaction to take appropriate action.

```
Method s.delete() raises(IntegrityError)
// Delete complex object s.
  t = Transaction::current();
  Transaction nt(t);
  nt.begin()
  try
    s._try_delete();
    nt.commit();
  handle IntegrityError
    nt.abort();
    raise IntegrityError;
  end handle
```

```
Method s._try_delete() raises(IntegrityError)
// Try to delete complex object s.
  t = Transaction::current();
  if t._deleted(s) then exit;
  t._mark_for_deletion(s);
  S = Type(s);
  for each traversal path tpR in S do
    for each target object r referenced by s.tpR do
      s._primitive_drop_tpR (r);
      tpS = Inverse(tpR);
      b._enforce_binding(ImpB(tpS), tpS);
    end for
  end for
  s._primitive_delete();
```

Fig. 12.8 Algorithms for the delete() and _try_delete() methods in interface Object

The _try_delete() method, given in Fig. 12.8, is a recursive method that may result in the implicit deletion of a number of objects that are related directly or indirectly to the object upon which it is invoked, designated here as s. Its invocation on an object must be dynamically bound to the method on the class representing the object's most specific type, i.e., it must be a virtual function in the C++ sense. This ensures that _try_delete() processes all traversal path instances involving the object.

The method first checks that object s has not already been marked for deletion by invoking the _deleted() method on the current transaction. If it has, _try_delete() simply exits. If not, it invokes _mark_for_deletion() on the current transaction to mark object s for deletion. The methods _deleted() and _mark_for_deletion() are given in Fig. 12.6.

The outer **for each** loop in Fig. 12.8 traverses every traversal path tp_R defined in class S, declared in the class or inherited. For each of these traversal paths in object s, the inner **for each** traverses all references in the traversal path. The purpose here is to attempt to implicitly drop each reference to a target object *r* (including the inverse reference to s) so that the object s can be deleted. The code first drops each such reference by invoking the _primitive_drop_tp_R method on s, which drops s.tp_R's reference to *r* and *r*.tp_S's reference to s. It then invokes the method _enforce_binding() on the target object *r* to enforce the implicit destructibility binding ImpB(tp_S) for the inverse traversal path tp_S. This method is presented in the next section.

The last step of _try_delete() actually deletes the object but is done only if all of the references to target objects could be dropped, i.e., only if none of the _enforce_binding() invocations resulted in an exception. The _primitive_delete() method is invoked on object s. It is assumed that this method simply deletes the primitive object from the database. It does not enforce referential integrity like the delete() method in the existing Object Model since this has already been done by _try_delete().

12.3.4 Method _enforce_binding()

The _enforce_binding() method, given in Fig. 12.9, is assumed for simplicity to be defined in the interface Object. It need only be defined for classes having relationships. The _enforce_binding() method for one class in a relationship must be accessible to the other class. The method enforces the destructibility binding semantics specified in Table 12.1. It is invoked from the _try_delete() method to enforce implicit destructibility bindings and from a drop_tp_R() and change_tp_R() method to enforce explicit destructibility bindings. Here, *r* denotes the implicit parameter and tp_S denotes an explicit parameter since _enforce_binding() is invoked on a target object to enforce the binding for the inverse traversal path in that target object. It is invoked after a reference to target object *r* and its inverse reference in the traversal path tp_S have just been dropped by the caller. The **case** statement executes the appropriate code for the given *binding*.

In the case of a nil (default) binding, the method _check_path_at_commit() is invoked on the current transaction when the lower bound multiplicity for the traversal path tp_S has been violated. _check_path_at_commit(), given in Fig. 12.6, inserts the traversal path instance into the transaction's *deferred_lbM_checks* set so that the lower bound constraint can be rechecked at, i.e., deferred to, the end of the complex object operation, which is at the end of the current, nested transaction.

For a default <association>, the nil case is always executed and the **if** condition is always false, resulting in nothing more than the already executed primitive drop with referential integrity enforced.

```
Method r._enforce_binding(binding binding, TraversalPath tpS) raises(IntegrityError)
// Enforce given binding for traversal path tpS in target object r.
  t = Transaction::current();
  case binding
    nil:  if Refs(r.tpS) < LbM(tpS) then  t._check_path_at_commit(r, tpS);
    - :  raise IntegrityError;
    ? :  if Refs(r.tpS) < LbM(tpS) then r._try_delete();
    ! :  r._try_delete();
    ' :  if Refs(r.tpS) < LbM(tpS) then  t._check_path_at_commit(r, tpS);
         Transaction nt(t);
         nt.begin()
         try
           r._try_delete();
           nt.commit();
         handle IntegrityError
           nt.abort();
         end handle
  end case
```

Fig. 12.9 Method _enforce_binding() in interface Object

The - binding case, i.e., a |- or X- binding, always results in an implicit exit with an IntegrityError exception. The relationship reference cannot be dropped.

The ? binding case invokes _try_delete() on the target object *r* when the lower bound multiplicity for the traversal path tp_S has been violated. If this invocation raises an exception, _enforce_binding() exits with an exception.

The ! binding case simply invokes _try_delete() on the target object *r*. If this invocation raises an exception, _enforce_binding() exits with an exception.

The ' binding case first invokes _check_path_at_commit() on the current transaction when the lower bound multiplicity for traversal path tp_S has been violated. The code then proceeds to invoke _try_delete() on the target object; however, unlike the ? or ! case, it invokes this method within another nested transaction so that the delete can be undone if an exception results. Here, such an exception is not re-raised. The implicit delete that was tried and failed is simply forgotten, which implements the unique, useful, and often applicable semantics of the ' binding. If the delete was required because of a lower bound multiplicity violation, an exception results at the commit of the nested transaction that ends the complex object operation (based on the prior call to _check_path_at_commit()).

It should be noted that before the _try_delete() method is invoked to implicitly delete a target object (for cases ?, !, and '), the reference to the target object and its inverse have already been deleted. Thus, this reference (or association instance) is not subsequently considered when determining whether or not an implicit delete of the target object is possible.

It should also be noted that link cycles in the database, in which an object references itself directly or indirectly, do not cause circularity within the given algorithms. The cascading deletes are guaranteed to terminate because of the "**if** *t*._deleted(*s*) **then exit**" statement in the _try_delete() method, which does a depth-first traversal of the database tracing out a tree of implicitly deleted objects.

12.3.5 Method form_tp_R()

The form_tp_R() method, given in Fig. 12.10, is generated by the ODMS for each traversal path, tp_R, defined in ODL. It is a method on a class *S* (see Table 12.1) that contains the traversal path tp_R. The method wraps the form_tp_R() method as currently defined in the Object Model. It provides the ORN semantics for the creation of an association link, i.e., the forming of a traversal path reference (including its inverse reference), yet retains the semantics of the currently defined form_tp_R() method when the traversal path has the default <*association*>. The form_tp_R method, like the delete() method, provides a nested transaction that embeds the complex object operation, making it atomic.

```
Method s.form_tpR(Object r) raises(IntegrityError)
// Form a relationship reference to object r in traversal path tpR of complex object s.
  t = Transaction::current();
  Transaction nt(t);
  nt.begin()
  try
    tpS = Inverse(tpR);
    if Refs(s.tpR) = UbM(tpR) or Refs(r.tpS) = UbM(tpS) then raise IntegrityError;
    s._primitive_form_tpR(r);
    nt.commit();
  handle IntegrityError
    nt.abort();
    raise IntegrityError;
  end handle
```

Fig. 12.10 Algorithm for the form_tp_R() method in class *S*

The method first checks that upper bound multiplicities for traversal path tp_R and its inverse tp_S will not be violated by forming the new relationship reference. For simplicity, it is assumed that the = operator returns false for upper bound multiplicity *. If a multiplicity is violated, an exception is raised. If not, the method invokes the primitive version of form_tp_R() on *s*, i.e., *s*._primitive_form_tp_R(*r*), which just adds to *s*.tp_R a reference to *r* and adds to *r*.tp_S a reference to *s*. The primitive method should raise an IntegrityError if a traversal path already contains the reference being added.

12.3.6 Method drop_tp_R()

The drop_tp_R() method, given in Fig. 12.11, is generated by the ODMS for each traversal path, tp_R, defined in ODL. It is a method on a class S (see Table 12.1) that contains the traversal path tp_R. The method wraps the drop_tp_R() method as currently defined in the Object Model. It provides the ORN semantics for the explicit destruction of an association link, i.e., the dropping of a traversal path reference (including its inverse reference), yet retains the semantics of the currently defined drop_tp_R() method when the traversal path has the default <association>. The drop_tp_R method provides a nested transaction that embeds the complex object operation, making it atomic.

```
Method s.drop_tpR(Object r) raises(IntegrityError)
// Drop the relationship reference to object r in traversal path tpR of complex object s.
  t = Transaction::current();
  Transaction nt(t);
  nt.begin()
  try
    s._primitive_drop_tpR (r);
    tpS = Inverse(tpR);
    r._enforce_explicit_binding(tpS);  // Enforces explicit binding for tpS with r as the target
                                       //   object
    s._enforce_explicit_binding(tpR);  // Enforces explicit binding for tpR with s as the target
                                       //   object
    nt.commit();
  handle IntegrityError
    nt.abort();
    raise IntegrityError;
  end handle
```

Fig. 12.11 Algorithm for the drop_tp_R() method in class S

The method first invokes the primitive version of drop_tp_R() on s, specifically s._primitive_drop_tp_R(r), which simply removes from s.tp_R the reference to r and removes from r.tp_S the reference to s. The primitive method should raise an IntegrityError if the traversal paths do not contain these references. If no exception is raised, drop_tp_R() proceeds to enforce the explicit destructibility bindings for each traversal path in the relationship, first the binding for tp_S (i.e., for class S) with r serving as the target object and then the binding for tp_R (i.e., class R) with s serving as the target object.

12.3.7 Method _enforce_explicit_binding()

The _enforce_explicit_binding() method, given in Fig. 12.12, is assumed for simplicity to be defined in the interface Object. Like _enforce_binding(), it need only be defined for classes having relationships. The method for one class in a relationship must be made accessible to the other class. (In OR+, this and the _enforce_binding() method are actually implemented in C++ on a base class for all traversal paths and inherited by the derived classes that are generated by the system for each traversal path. The class for a traversal path and that for its inverse traversal path are declared as friends.)

```
Method r._enforce_explicit_binding(TraversalPath tpS) raises(IntegrityError)
// Enforce explicit binding for traversal path tpS in target object r.
  t = Transaction::current();
  if t.in_rxc_mode() and ExpB(tpS) ≠ – then
    if Refs(r.tpS) < LbM(tpS) then t._check_path_at_parent_commit(r, tpS);
    exit;
  end if
  r._enforce_binding(ExpB(tpS), tpS);
```

Fig. 12.12 Algorithm for the _enforce_explicit_binding() method in interface Object

The _enforce_explicit_binding() method first handles the situation when RXC mode is set and the explicit destructibility binding for the traversal path is not -, i.e., not X-. (This binding subsequently generates an exception.) If the lower bound multiplicity for the traversal path is violated, the _check_path_at_parent_commit() method, given in Fig. 12.6, is invoked on the current transaction. The traversal path instance is essentially passed to this method, which inserts it into the *deferred_lbM_checks* set of the parent transaction, which is the user-defined transaction that invoked drop_tp_R().

If RXC mode is not set or the binding for the traversal path is X-, the method invokes the _enforce_binding() method on the target object *r* to enforce the explicit destructibility binding ExpB(tp_S) for the inverse traversal path tp_S. This method was presented in Section 12.3.4.

12.3.6 Method change_tp_R()

The change_tp_R() method, given in Fig. 12.13, is generated by the ODMS for each traversal path, tp_R, defined in ODL. It is a method on a class *S* (see Table 12.1) that contains the traversal path tp_R. The method does not have an equivalent method in the Object Model but uses the form_tp_R() and drop_tp_R() methods as currently defined in the model. It provides the ORN semantics for the explicit change of an association link, replacing one related object reference with another. The change_tp_R method provides a nested transaction that embeds the complex object operation, making it atomic.

Within the nested transaction, the method first does the "drop" portion of the relationship change operation. It invokes the primitive version of drop_tp_R() on *s*, specifically *s*._primitive_drop_tp_R(*r1*), which simply removes from *s*.tp_R the reference to *r1* and removes from *r1*.tp_S the reference to *s*. This method should raise an IntegrityError if the traversal paths do not contain these references. If no exception is raised, the method invokes the _enforce_explicit_binding() method, presented in the previous section, to enforce the explicit destructibility binding for tp_S (i.e., for class *S*) with *r1* serving as the target object. The explicit destructibility binding for tp_R is not enforced. (See footnote 1 for Table 12.1.)

```
Method s.change_tpR(Object r1, Object r2) raises(IntegrityError)
// Change traversal path tpR of complex object s to reference object r2 instead of object r1.
  t = Transaction::current();
  Transaction nt(t);
  nt.begin()
  try
    s._primitive_drop_tpR (r1);
    tpS = Inverse(tpR);
    r1._enforce_explicit_binding(tpS);  // enforces explicit binding for tpS with r1 as the target
                                        //    object
    if t.deleted_objects.contains_element(s) or t.deleted_objects.contains_element(r2) then
      raise IntegrityError;
    if Refs(r2.tpS) = UbM(tpS) then raise IntegrityError;
    s._primitive_form_tpR(r2);
    nt.commit();
  handle IntegrityError
    nt.abort();
    raise IntegrityError;
  end handle
```

Fig. 12.13 Algorithm for the change_tp_R() method in class *S*

The change_tp_R() method next does the "form" portion of the relationship change operation. It first checks that neither objects s nor *r2* have been implicitly deleted as a result of enforcing the tp_S binding. Then it checks that the upper bound multiplicity for traversal path tp_S will not be violated by forming a new relationship reference between object *r2* and object s. Again, for simplicity, it is assumed that the = operator returns false for upper bound multiplicity *. If a multiplicity is violated, an exception is raised. If not, the method invokes the primitive version of form_tp_R() on s, i.e., s._primitive_form_tp_R(*r2*), which just adds to s.tp_R a reference to *r2* and adds to *r2*.tp_S a reference to s. The primitive method should raise an IntegrityError if a traversal path already contains the reference being added.

12.3.9 Example

To aid in the reader's understanding of the algorithms presented here, we will trace an example execution of delete() to illustrate the enforcement of ORN semantics and the recursive nature of the implementation. Let us assume the ODL in Fig. 12.5, the sample company database shown in Fig. 12.14, and the invocation of e2.delete() to delete employee e2. Fig. 12.14 is the same as Fig. 3.7. The class diagram and database was created using the ORN Simulator.

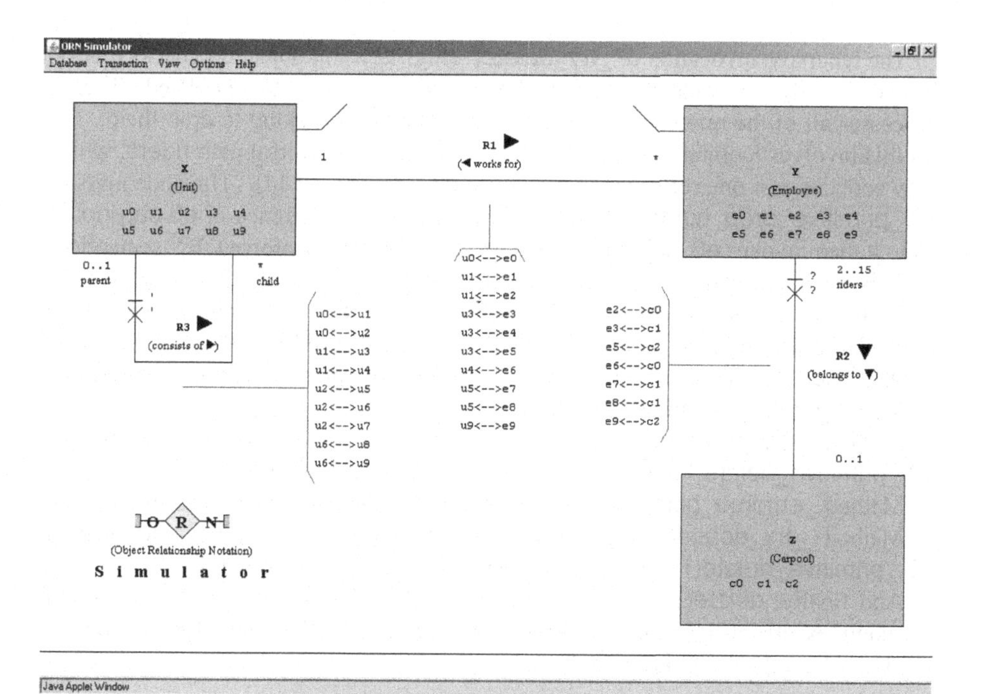

Fig. 12.14 Sample company database

The method delete() (Fig. 12.8) invokes e2._try_delete() within the nested transaction *nt*. Assume that the type of e2 is actually SalesPerson (a subclass of Employee not shown in Fig. 12.14). In this case the _try_delete() method for the class Salesperson is executed.

The method _try_delete() (Fig. 12.8) checks that e2 has not already been deleted (it has not) and marks the object to be deleted. Then, it processes all of the references from e2 to related objects seeking to drop them. For e2 this involves looping through all of the references in traversal path unit and carpool (both shown in Fig. 12.5) and perhaps other traversal paths, e.g., orders, that are not shown in Fig. 12.5 or Fig. 12.14. e2.unit contains a reference to unit u1, and e2.orders may contain many references to the orders that employee e2, as salesperson e2, has obtained. To focus on the carpool traversal path, we assume that the unit reference to u1 is successfully dropped (it will be since the implicit destructibility binding for Employee is default and the lower bound multiplicity is 0) and that any references in a traversal path orders are also successfully dropped For the reference in carpool to c0, the code invokes e2._primitive_drop_carpool(c0) to drop it (along with the reference in c0.riders to e2). It then invokes c0._enforce_binding(**?**, riders) to enforce ORN semantics.

The _enforce_binding() method (Fig. 12.9) executes the code for case **?**, which is: **if** Refs(c0.riders) < LbM(riders) **then** c0._try_delete(). Since the value of Refs(c0.riders) is now 1 and LbM(riders) is 2, an implicit delete of c0 is attempted.

The recursive invocation of _try_delete() (Fig. 12.8) on object c0 checks that c0 has not already been deleted (it has not) and marks the object to be deleted. Then it processes all of the references from c0 to related objects seeking to drop them. For c0 this involves looping through all of the references in traversal path riders, which now contains just one reference to employee e6 (see Fig. 12.14). The code invokes c0._primitive_drop_riders(e6) to drop it (and also the reference in e6.carpool to c0). It then invokes e6._enforce_binding(nil, carpool) to enforce ORN semantics.

This recursive call to the _enforce_binding() method (Fig. 12.9) executes the code for case nil, which is: **if** Refs(e6.carpool) < LbM(carpool) **then** *t*._check_path_at_commit(e6, carpool). Since LbM(carpool) is 0 (and Refs(e6.carpool) is now 0 but irrelevant), the **then** statement is not executed and _enforce_binding() exits successfully.

Method _try_delete() invoked on carpool c0 now invokes the method c0._primitive_delete() to actually delete c0 and then exits successfully.

Method _enforce_binding() invoked on carpool c0 now exits successfully.

Method _try_delete() invoked on employee e2 now invokes the method e2._primitive_delete() to actually delete e2 and then exits successfully.

And finally, delete() invoked on employee e2 invokes *nt*.commit() to commit the complex object operation. No exceptions are raised, the method exits, and the deletion of employee e2 is successful.

12.4 Conclusion

This chapter has shown how ORN can be added to the ODMG Object Model and has presented and discussed algorithms for implementing ORN semantics in an ODMS. The shortcomings of adding ORN to the Object Model are a slight increase in complexity and a requirement that ODMS implementations include a nested transaction capability for ORN implementation. Despite these shortcomings and regardless of whether or not ORN is added to any standard for ODMS, ODMS vendors should consider including ORN as an extended feature to their system because:

- ORN is a simple notation that allows the database developer to specify a large variety of association semantics, which define the scopes of complex and composite objects.
- The extended ODL would facilitate a straightforward mapping of association semantics from a conceptual database model, expressed as an ORN-extended UML class diagram, to the logical database model, expressed in the ODL.
- The ODMS would provide the same support for associations that is provided by relational DBMSs via the SQL references clause plus support even more powerful association semantics.
- If no *<association>* is given for a traversal path, the default *<association>* corresponds to current system capabilities. Thus, adding ORN is a pure extension requiring no changes to the underlying Object Model capabilities.

- The implementation of this extension is relatively simple as shown by the algorithms given here and their successful implementation in OR+.
- The benefits are increased database development productivity and improved database integrity as much less code needs to be developed and maintained by database application developers.

Bibliography

Albano A, Ghelli G, Orsini B (1991) A relationship mechanism for a strongly typed object-oriented database programming language. *Proc VLDB Conf,* Morgan Kaufmann, 565-575

Albert M, Pelechano V, Fons J, Ruiz M, Pastor O (2003) Implementing UML association, aggregation, and composition: a particular interpretation based on a multidimensional framework. *Proc CAiSE 2003 Conf,* Eder J and Missikoff M (eds), Springer-Verlag LNCS 2681:143-158

ANSI (2008) Information technology - database languages - SQL - part 2: foundation (SQL/Foundation), ISO/IEC 9075-2:2008. American National Standards Institute (ANSI), New York, www.webstore.ansi.org. Accessed 4 March 2009

ANSI (1975) ANSI/X3/SPARC Study Group on Database Management Systems, Interim Report, FDT. ACM SIGMOD Bulletin 7, 2.

Balaban M, Shoval P (2000) MEER – A EER model enhanced with structure methods. Inf Syst 27, 4:245-275

Barbier F, Henderson-Sellers B, Parc-Lacayrelle A, Bruel J (2003) Formalization of the Whole-Part Relationship in the Unified Modeling Language. IEEE Trans on Softw Eng 29, 5:459-469

Bertino E, Martino L (1991) Object-oriented database management systems: concepts and issues. IEEE Computer 24, 4:33-47

Booch G, Rumbaugh J, Jacobson I (1999a). The Unified Modeling Language user guide. Addison Wesley, Reading, MA

Booch G, Rumbaugh J, Jacobson I (1999b). The Unified Modeling Language reference manual. Addison-Wesley, Reading, MA

Bouzeghoub M, Metais E (1991) Semantic modeling and object oriented databases. *Proc VLDB Conference*, Morgan Kaufmann, 3-14

Brown SG (1997) The ORN Simulator: a practical tool for modeling relationship behavior. *Proc ADMI Conf,* 130-135

Cattel RGG, Barry DK, Berler M, Eastman J, Jordan D, Russell C, Schadow O, Stanienda T, Velez F (2000) The object database standard: ODMG 3.0. Morgan Kaufmann, San Mateo, CA

Cattel RGG, et al. (1997) The object database standard: ODMG 2.0. Morgan Kaufmann, San Francisco, CA (see Release 1.1 for description of future binding)

Chen PP (1976) The entity-relationship model: towards a unified view of data. ACM Trans on Database Syst 1, 1:1-36

Civello F (1993) Roles for composite objects in object-oriented analysis and design. *Proc OOPSLA Conf,* ACM, 376-393

Codd EF (1970) A relational model of data for large shared data banks. Commun of the ACM 13, 6:377-387

Codd EF (1979) Extending database relations to capture more meaning. ACM Trans on Database Syst 4, 4:397-434

Codd EF (1990) The relational model for database management—version 2. Addison-Wesley, Reading MA

Connolly T, Carolyn B (2005) Database systems: a practical approach to design, implementation, and management. Addison-Wesley, Reading, MA, 467

Copeland G, Maier D (1984) Making SmallTalk a database system. *Proc SIGMOD Conf,* ACM, 316-325.

Date CJ (1981) Referential Integrity. *Proc VLDB Conf,* 2-12

B.K. Ehlmann, *Object Relationship Notation (ORN) for Database Applications,*
Advances in Database Systems 39, DOI 10.1007/978-0-387-09554-7_BM2,

Date CJ (1990) Relational database writings 1985-1989. Addison-Wesley, Reading, MA, 119-125, 143-147

Date CJ, Darwen H (1994) A guide to the SQL standard, third edition. Addison-Wesley, Reading, MA, 399-401

Ehlmann B (2008) Adding more support for associations to the ODMG Object Model. *ICSOFT 2006 Conf: Revised Selected Papers*, Softw and Data Technol, Springer-Verlag, Filipe J, Shishkov B, and Helfert M (eds), CCIS 10:257-269

Ehlmann BK (2007) ORN Additive: shrinking the gap between database modeling and implementation. *Proc ICIS Conf*, IEEE Computer Society, 555-560

Ehlmann BK (2006) Incorporating Object Relationship Notation (ORN) into SQL—revisited. *Proc ACM Southeast Conf*, 389-394

Ehlmann BK (2002) A data modeling tool where associations come alive. *Proc IASTED MIC Conf*, 66-72

Ehlmann BK (1992) Applying an object-oriented database model to a scientific database problem: managing experimental data at CEBAF. Ph.D. dissertation, UMI Dissertation Services, Ann Arbor, MI

Ehlmann BK, Riccardi GA (1994) A notation for describing aggregate relationships in an object-oriented data model. Springer-Verlag Lect Notes in Comput Sci 819:62-77

Ehlmann BK, Riccardi GA (1996) A comparison of ORN to other declarative schemes for specifying relationship semantics. Inf and Softw Technol 38, 7:455-465

Ehlmann BK, Riccardi GA (1997a) An integrated and enhanced methodology for modeling and implementing object relationships. J of Object-Oriented Program 10, 2:47-55

Ehlmann BK, Riccardi GA (1997b) Object Relater *Plus*: a practical tool for developing enhanced object databases. *Proc Data Eng Conf*, IEEE Computer Society Press, 412-421

Ehlmann BK, Riccardi GA (1999) Object Relationship Notation (ORN) and the ORN Simulator. *Demos and Posters Proc ER Conference*, 9-10

Ehlmann BK, Stewart MA (1997) Incorporating Object Relationship Notation (ORN) into SQL. *Proc ACM Southeast Conf*, 282-289

Ehlmann BK, Yu X (2005a) Generating SQL to implement enhanced association semantics. *Proc WorldComp IKE Conf*, 120-127

Ehlmann BK, Yu X (2005b) The difficulty of mapping modeled associations to SQL. *Proc IASTED DBA Conf*, 65-70

Ehlmann BK, Yu X (2002) Extending UML class diagrams to capture additional association semantics. *Proc IASTED Applied Informatics Conf*, 395-401

Ehlmann BK, Riccardi GA, Dennis LC (1992) Representing non-inheritance relationships in an object-oriented, scientific database. *Proc SSDBM Conf*, ETH Zurich, 99-109

Ehlmann BK, Dennis LC, Riccardi GA (1993) An object-based conceptual model of a nuclear physics experiments database. Nucl Instrum & Methods in Phys Res, Sect A A325,1&2:294-308

Ehlmann BK, Rishe N, Shi J (2000) The formal specification of ORN semantics. Inf and Softw Technol 42, 3:159-170

Ehlmann BK, Riccardi GA, Rishe N, Shi J (2002) Specifying and enforcing association semantics via ORN in the presence of association cycles. IEEE Trans on Knowl and Data Eng 14, 6:1249-1257

Embley DW, Ling TW (1990) Synergistic database design with an extended entity-relationship model. In: Lochovsky FH (ed) Entity-relationship approach to database design and query. North Holland, New York

Fowler M (1997) Analysis Patterns: Reusable Object Models, Addison-Wesley, Reading, MA

Gamma E, Helm R, Johnson R, Vlissides, J (1995) Design patterns: elements of reusable object-oriented software. Addison-Wesley, Reading, MA

Guo M, Su SYW, Lam H (1991) An association algebra for processing object-oriented databases. *Proc Data Eng Conf*, 23-32

Hardeman SK, Ehlmann BK (1996) Relationship behavior in object databases: subtleties and inconsistencies. *Proc ACM Southeast Conf*, 224-229

Hay DC (1996) Data model patterns: conventions of thought. Dorset House, New York

Henderson-Sellers B, Barbier F ((1999) What is this thing called aggregation? *Proc of TOOLS 29*, IEEE Computer Society, 216-230.

Horowitz BM (1992) A run-time execution model for referential integrity maintenance. *Proc Data Eng Conf*, 548-556

Kilov H, Ross J (1994) Information modeling: an object-oriented approach. Prentice Hall, Englewood Cliffs, NJ, 268

Kim W (1990) Object-oriented databases: definition and research directions. IEEE Trans on Knowl and Data Eng 2, 3:327-341

Kim W, Bertino E, Garza JF (1989) Composite objects revisited. *Proc ACM SIGMOD Conf.* In: ACM SIGMOD RECORD 18, 2:337-347

Kolp M, Zimanyi E (2000) Enhanced ER to relational mapping and interrelational normalization. Inf and Softw Technol 42:1057-1073.

Lazarevic B, Misic V (1991) Extending the entity-relationship model to capture dynamic behavior. European J of Inf Syst 1, 2:95-106

Markowitz VM (1990) Referential integrity revisited: an object-oriented perspective. *Proc VLDB Conf*, Morgan Kaufmann, 578-589

Markovitz M, Shoshani A (1992) Representing extended entity-relationship structures in relational databases: a modular approach. ACM Trans on Database Syst 17, 3:423-464

Mellor SJ, Clark AN, Futagami T (2003) Guest editor's introduction: model-driven development. IEEE Softw 20, 5:19-25

Melton J (2003). Advanced SQL: 1999—Understanding Object-Relational and Other Advanced Features. Morgan Kaufmann, San Francisco, CA, 119-151.

Micosoft Inc. (2008) Microsoft SQL Server 2008. www.microsoft.com/sql/default.mspx Accessed 27 February 2009.

Neal E, Ehlmann BK (2000) A new UML-compatible object relationship notation (ORN). *Proc ACM Southeast Conf*, 179-183

Odell JJ (1994) Six different kinds of composition. J of Object-Oriented Program 5, 8:10-15

OMG (2005) OMG unified modeling language specification, version 2.0. Object Management Group, www.omg.org. Accessed 27 February 2009.

ONTOS (1994) ONTOS DB 3.0 developer's guide. ONTOS Inc, Burlington, MA

Progress Software Inc (2006) ObjectStore interprise. www.objectstore.com/datasheet/index.ssp Accessed 27 February 2009.

Ricardo CM (2004) Databases Illuminated. Jones and Bartlett, Sudbury, MA

Riccardi GA, Ehlmann BK (1991) Object-oriented development of scientific databases, an example from experimental physics. *Proc SERF Conf*, Rodriguez RV (ed), 277-286

Rumbaugh J (1987) Relations as semantic constructs in an object-oriented language. *Proc OOPSLA Conf*, ACM, 466-481

Rumbaugh J (1988) Controlling propagation of operations using attributes on relations. *Proc OOPSLA Conf*, ACM, 285-296

Rumbaugh J, Blaha M, Premerlani W, Eddy F, Lorensen W (1991) Object-oriented modeling and design. Prentice Hall, Englewood Cliffs, NJ

Rundensteiner EA, Bic L, Gilbert JP, Yin M (1994) Set restrictions for semantic groupings. IEEE Trans on Knowl and Data Eng 6, 2:193-204

Saksena M, France RB, Larrondo-Petrie (1998) A characterization of aggregation. *Proc OOIS Conf*, Springer, Roland C and Grosz G (eds), 11-19

Shipman DW (1981) The functional data model and the data language DAPLEX. ACM Trans on Database Syst 6, 1:140-173

Smith J, Smith D (1977) Database abstractions: Aggregation and Generalization. ACM Trans on Database Syst 2, 2:105-133

Snoeck M, Dedene G (2001) Core modeling concepts to define aggregation. L'Object 7, 3:281-306

Stonebraker M, Rowe LA, Hirohama M (1990) The implementation of POSTGRES. IEEE Trans on Knowl & Data Eng 2, 1:125-142

Tan P, Steinbach M, Kumar V (2006) Introduction to data mining. Addison-Wesley, Reading, MA

Teorey T (1990) Database modeling and design: the Entity-Relationship approach. Morgan Kaufmann, Los Altos, CA

Tsichritzis DC, Lochovsky FH (1982) Data models. Prentice-Hall, Englewood Cliffs, NJ

Ullman J (1982) Principles of database systems. Computer Science Press, Rockville, Maryland

VERSANT (1993) VERSANT ODBMS system reference manual release 3.0 VERSANT Object Technologies, Menlo Park, CA

Winston M, Chaffin R, Herrmann (1987) A taxonomy of part-whole relations. Cognitive Science 11: 417-444

Wordsworth JB (1992) Software development with Z. Addison-Wesley, Wokingham, UK

Zaniolo C (1983) The Database Language GEM. *Proc SIGMOD Conf*, ACM, 207-218

Zdonik SB, Maier D (1990) Fundamentals of object-oriented databases. In: Zdonik SB, Maier D (eds) Readings in object-oriented DB systems. Morgan Kaufmann, San Mateo, CA

Index